Y0-CDJ-950

INFORMATION SYSTEMS ARCHITECTURE:

A System Developer's Primer

OTHER INMON BOOKS

Design Review Methodology for a Data Base Environment (written in association with others)

Effective Data Base Design

Integrating Data Processing Systems: In Theory and in Practice

Management Control of Data Processing: Preventing Management-By-Crisis

INFORMATION SYSTEMS ARCHITECTURE:

A System Developer's Primer

W. H. INMON
Coopers & Lybrand

PRENTICE-HALL, INC., Englewood Cliffs, NJ 07632

Library of Congress Cataloging-in-Publication Data

INMON, WILLIAM H.
Information systems architecture.

Bibliography: p. 229
Includes index.
1. System design. 2. System analysis.
3. Management information systems. I. Title.
QA76.9.S88I55 1986 658.4'03'2 85-12051
ISBN 0-13-464694-0

Editorial/production supervision and
interior design: Tom Aloisi
Cover design: Diane Saxe
Manufacturing buyer: Gordon Osbourne
Page layout: Diane Koromhas

Printed in the United States of America

10 9 8 7 6 5 4 3 2

ISBN: 0-13-464694-0 01

PRENTICE-HALL INTERNATIONAL, (UK) LIMITED, *London*
PRENTICE-HALL OF AUSTRALIA PTY. LIMITED, *Sydney*
EDITORA PRENTICE-HALL DO BRASIL, LTDA., *Rio de Janeiro*
PRENTICE-HALL CANADA INC., *Toronto*
PRENTICE-HALL HISPANOAMERICANA, S.A., *Mexico*
PRENTICE-HALL OF INDIA PRIVATE LIMITED, *New Delhi*
PRENTICE-HALL OF JAPAN, INC., *Tokyo*
PRENTICE-HALL OF SOUTHEAST ASIA PTE. LTD., *Singapore*
WHITEHALL BOOKS LIMITED, *Wellington, New Zealand*

— For my sister, Kay Ann

Contents

Preface

This book addresses the heart of information systems architecture—the modeling of the data and processes of a company and how that model relates to the business of the company. Whether a shop is building operational or decision-support systems, the *business* of the company dominates. Modeling data and processes is the means by which the business manager translates his or her needs to the computer person. Without a stable model, the resulting data processing system that is being built to serve the needs of the business will be based on unstable ground.

This book can be compared to two books: *Structured Systems Analysis* by C. Gane and T. Sarson (Prentice-Hall, 1979) and *Data Models* by D. C. Tschritzsis and F. H. Lochovosky (Prentice-Hall, 1981). This book represents the middle ground represented by these two books. The two aforementioned books address one side of the coin of data and processes, but not both, while this books brings both together. There is a balance placed on the modeling of both data and processes.

This book is aimed at the most basic level of reader. It is intentionally short on theory and long on practicality. The earlier book from which it stems, *Integrating Data Processing Systems: In Theory and in Practice* (Prentice-Hall, 1984), serves as a theoretical foundation for nearly all of the theory found in this book. What little theory that is introduced is reduced to the most basic level and is explained in great detail and with examples. An idea is explained, then described in terms of at least one example. It is anticipated that the reader can use this book as a step-by-step guide to the creation of a business-based system model. Throughout the book three major models (over different points in the modeling process) are developed: one for a financial system, one for a manufacturing system, and one for insurance systems.

The most interested industrial audience will be from data administration and data base administration. Other interested parties will be designers, developers, and managers. From the academic audience it is anticipated that students of computer science will find the book attractive. It is aimed at the first course in system modeling, probably at the senior or graduate level. The size of the shop and the vendor, the languages used, the DBMS used, will have no bearing on the level of interest of the audience. In addition, the shelf life of the book will be quite lengthy, as the concepts discussed will not change readily.

At the foundation of business-based modeling is a fundamental understanding of the business of the systems that run and are used to manage a company or organization. The data and processes that are at the heart of a company must be understood in their most natural or basic state if a stable foundation is to be built. Without a stable foundation the benefits of information engineering cannot be realized. The first step toward a stable systems foundation is an understanding of the form, function, and interrelationship of the components of the foundation. When data and processes have been modeled at the appropriate level and with the right techniques, the organization is prepared to build systems that will stand the test of time.

The common models (both data and processes) that are found throughout the book are developed from the high level conceptual models (i.e., the entity relationship model) to the point of models ready to be molded into a physical form (i.e., ready to be put into a database). Processes are decomposed from the highest level to the point where program specifications are able to be written. The reader is able to follow the progression in a cookbook fashion and is led to see how the techniques and considerations of modeling apply to his or her environment. The modeling techniques (and the reasons for doing modeling) are basic to the data processing environment, whether a shop is building operational, decision support, or archival systems. In all environments the resultant system is based on the *business* of the company and therein lies the usefulness of modeling the data and processes of the company.

In many ways this book is an extension of *Integrating Data Processing Systems: In Theory and In Practice,* whose trust is a description of many aspects and considerations of the integrated environment and a brief outline of how to achieve and maintain an integrated environment. This book takes one of the subjects—the modeling of data and processes—and greatly elaborates. In addition, this book is written at a *most basic level,* so that the complexities and mysteries of *what* modeling is and *why* modeling is essential are revealed.

This book reflects a change in philosophy that might escape the reader at first glance. In the early days of structured design and analysis, there was very heavy analytical emphasis on the processes of functions being done by a company, with little or no emphasis on the data. Data was considered to be a by-product of processing. Then, as if a pendulum had swung, there appeared works that heavily emphasized the data that a company runs on, almost to the exclusion of processing. The pendulum had swung back the other way. This book balances the emphasis on data and processes and in doing so, balances the pendulum between the two. The underlying assumption is that data and processes are interdependent and inseparable, like yin and yang.

A final underlying concept is that, to be successful, data and process modeling must be *simple.* At all stages of modeling, there is an emphasis on simplicity and at the same time, relevancy and accuracy. The modeling of data and processes has an almost infinite number of combinations, each of which has some degree of merit. Therefore, to be successful, the system modeler must cut through the complexity without losing relevancy and accuracy. This book illustrates at a most basic level techniques and considerations that can be used to create useful, simple, relevant, and accurate models of data and processes.

After selected chapters there is a project study, based on a fictional bank, but illustrating very nonfictional problems. The project study puts the concepts of modeling discussed in the chapter into a context of reality. As the reader studies the concepts in each chapter there may be some confusion as to exactly what is being discussed and the order in which activities occur. The project study allows the reader to put the concepts to work. Upon reading all the project studies, the reader is exposed to all the steps in business-based modeling. The project study leads the reader through all the steps in the construction of a model. While a complete business-based system model is too large to be constructed and placed in this book, there is a detailed example of the building of at least some part of the model of each step in the project study so that the reader can see an example of the output of each step.

ACKNOWLEDGEMENTS

A while ago Julian Tippett, British Airways, asked me some questions about data analysis, and at the time I gave him less than a satisfactory answer. I hope this book suffices to answer Julian's questions, even if quite a bit of time has passed.

Several people have been instrumental in the writing of this book. Indirectly, Bob Brown has influenced the material through his pioneering work. Jeanne Friedman has also contributed through her many conversations. Coopers and Lybrand has been supportive throughout. And finally, Melba Inmon has made countless small, medium, and large contributions in many ways. Without her continuing support this book would still be wishful thinking.

W. H. Inmon

INFORMATION SYSTEMS ARCHITECTURE:

A System Developer's Primer

1

Introduction: Terms, Tools, and Techniques

AN OVERVIEW

Systems that are centered around the basic business of a company are more stable and elastic than systems based around a piecemeal collection of requirements. Business-based systems are developed once and are maintained once, as opposed to piecemeal systems that are developed many times and maintained many more times. Decision-support systems that are based on the business of the company yield accurate and meaningful results as opposed to decision-support systems that are based on piecemeal development. The result of basing decision-support systems on piecemeal development is information that is untimely, unsynchronized, and often biased.

At the heart of the business-based systems is a model of the business—its data and its processes—that can be used equally by the data processing professional and the non-data processing professional as grounds for communication. The model will change *only* as the nature of the business changes, and therein lies the primary reason why business-based systems are stable. Systems developed on a piecemeal basis change every time some small aspect of the business changes, and the impact of change is felt in many places. But building a model properly and using it effectively are complex activities, because a practically infinite number of models can be constructed—each of which has some legitimate basis! The primary difficulty with systems modeling is that it appears to be very easy to do. What is difficult to do is build an *effective* system model. To be effective a model must:

- Communicate to the non-data processing community
- Communicate to the data processing community

- Define the data of a system uniquely, completely, simply, and in an organized manner representative of the usage of the data throughout the business
- Define the processes of a system at both the highest levels and lowest levels
- Represent the business of the company in its most basic form

Ironically, the reasons for doing business-based modeling and knowing what to do with the model once built are often more important than how the model was built. If the designer knows *why* to model and *what to do* with the model once built, the techniques of modeling are relatively unimportant. Conversely, if a designer does not understand why to model or what to do with the model once built, the best modeling techniques that are available will be wasted.

The end result of a business-based approach to systems is a high degree of integration, where systems are tightly interwoven and there is no overlap. The result is much like a jigsaw puzzle that has been completed, where each part of the integrated system interlocks with its neighbors to form a cohesive picture (Fig. 1.1).

To build a business model of a company and then to translate that model into integrated systems requires a change in management approaches from that normally found in the data processing shop. The traditional management approach can be

Figure 1.1

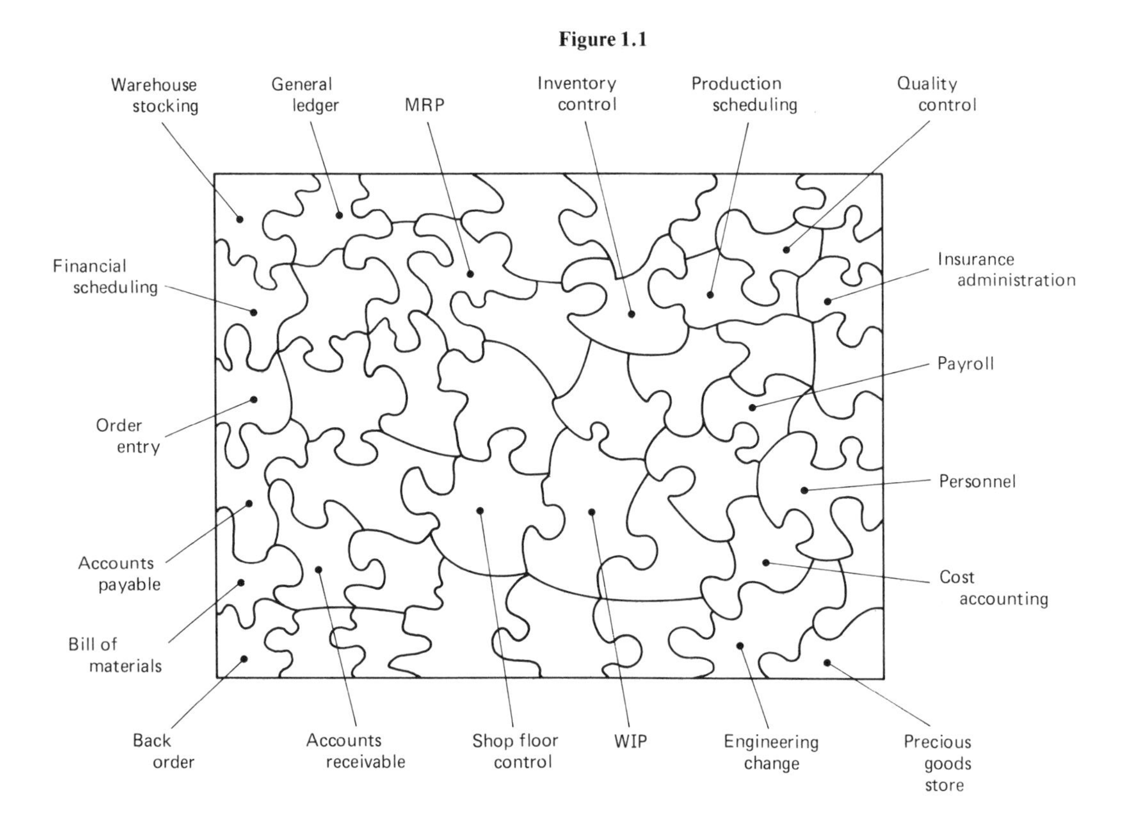

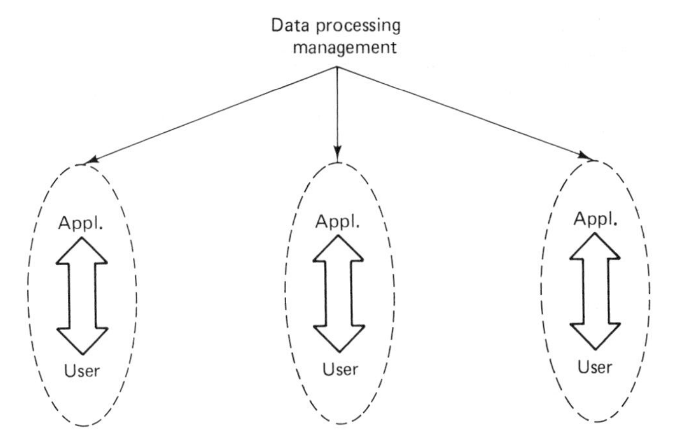

Figure 1.2 The local approach to management.

termed the *local* approach, because each user of data processing services is treated as if the user's immediate, or local, needs, independent of any other user, were the driving force in the company. The result of the local perspective of management spread over many users is many small, piecemeal systems being built, many of which duplicate in part or in total other systems. The local approach is shown in Fig. 1.2.

But as a shop grows and the organization matures, it becomes apparent that there are many important needs that must be met *across* the organization, or *globally*. Looking after each user individually does not satisfy many types of needs. Some needs are so large that all users and all systems are affected. One obvious global need is on-line system performance (i.e., on-line response time). Another global need is integration of data and processes across all systems. Figure 1.3 depicts the management approach required to address global problems.

The business of the company applies to all users. Therefore, it makes no sense to build models of the business for any one user in the absence of any other user. A business model includes the needs of *all* users, in much the same way as an on-line performance plan includes all on-line users. But a global approach to design or management is normally resisted because (1) it represents change and requires organiza-

Figure 1.3 The global approach to management.

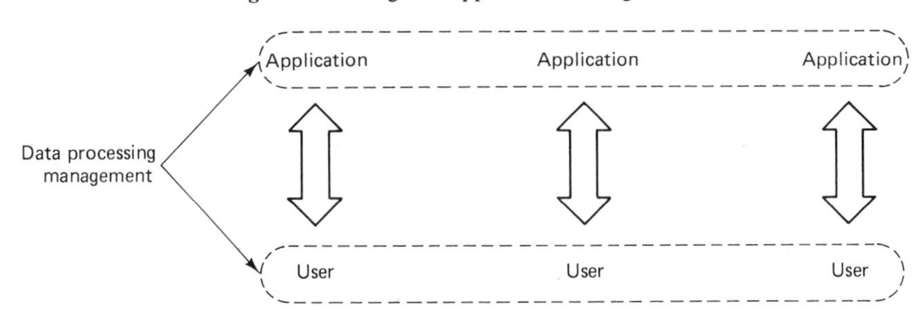

tional discipline, and (2) it often questions the domain of responsibility of a user. To put a global perspective into place requires resolve and discipline on the part of management. Without management understanding, support, and leadership, business-based systems are very difficult (if not impossible) to achieve.

BUSINESS-BASED MODEL CHARACTERISTICS

What are the general characteristics of a business-based system model? The data is modeled such that:

- Keys are defined globally. For example, in a bank a customer of one branch can be identified at another branch.
- Wholesale redundancy is reduced: Any data element is defined in a single place. For example, in a bank the address of a customer is not found with the data for his or her loans, savings, direct deposit accounts (DDAs), and so on, but in a single depository, the customer information file.
- There is compatibility at critical interfaces: Functionally different and yet related systems can communicate. For example, an overdrawn DDA can be noted by an officer when a loan application is being processed in a bank.
- Systems intermesh well: The volumes of data processed, hours of operation, the scope of activity, and so on, are compatible. For example, the on-line account post processing does not prevent daily activity balancing from occurring, even when there is an overlap in hours processed in a bank.

In short, data that is business-based provides a firm foundation on which future systems may be built and current systems run.

The model of processes in a business-based system have the following characteristics:

- A single process definition exists for processes common to multiple systems for each individual system. The programming code that calculates daily interest exists at the source code level in a single place (i.e., standard interest calculation algorithms do not exist separately throughout a bank's many systems).
- A single requirements definition exists for every major process. An inventory process for piece goods is not defined separately from an inventory process for raw materials or work in progress, for example.
- There is an ability to communicate from one process to another. As data is processed and values change, the results are known throughout all systems, not just the system doing the processing. For example, when a checking account is overdrawn, that information is readily available to the loan systems, the credit systems, the bankcard systems, and so on.

- Representation of the business processes exists at the most basic level. The processes most important and central to the business are the focus of the business-based model. For example, a business-based model for a bank would focus on transactions, loans, accounts, and deposits, as opposed to word processing, office automation, and inventory processing.

The main characteristics of a business-based model (both data and processes), then, are a minimum of redundancy within the model and representation of the business at the most basic level. In addition, the focus of the business-based model is on the definition of the system, not the implementation (e.g., to the data processor the embodiment of the successful business model will be at the source code and data definition level, not the object code and control block level). When a business-based model is properly defined, the implementation will naturally be highly integrated, but when a business-based model is not created, either formally or informally, no amount of effort at the implementation level will suffice to make the system integrated. Furthermore, once an unintegrated system is built, integration cannot be retrofitted after the fact.

The usefulness of the modeling process is best illustrated in terms of a data processing environment built without a business model. In the unmodeled environment, growth is haphazard, unplanned, overlapping, and generally disorganized. Further, the more growth there is, the more difficult the environment is to manage. This unmodeled, unintegrated growth is depicted symbolically in Fig. 1.4, where boxes represent systems being built. One system is built at day 1; another overlapping system built at day 2. By day 6 there are many overlapping, redundant, related systems.

Figure 1.4 Growth in the unmodeled environment.

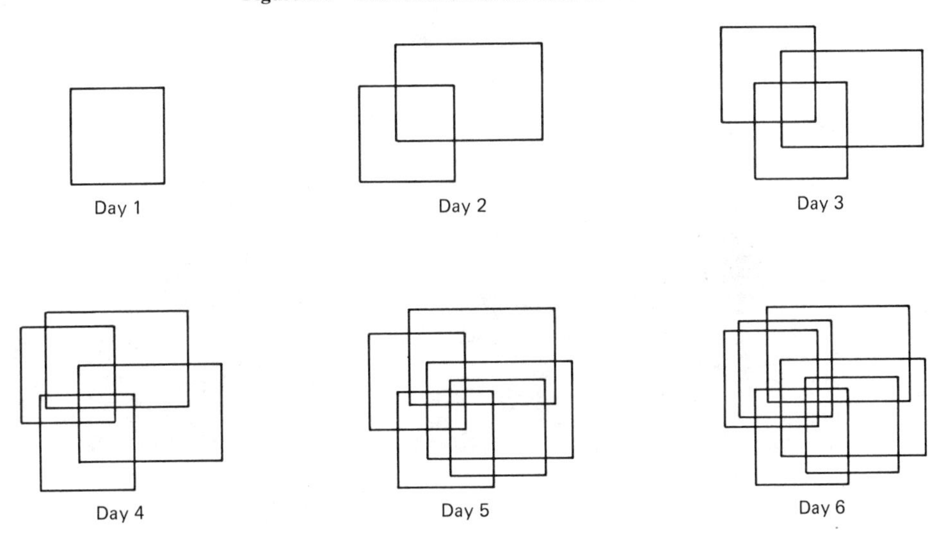

The opposite of unmodeled growth is planned, orderly growth, where interlocking systems are developed according to a blueprint, as in the case of business-based modelling. This growth is depicted by Fig. 1.5. The different stages of growth are depicted by the pieces of the jigsaw puzzle being filled in. Each piece connects with its neighbor and does not duplicate previous work done. The business-based system model serves as a blueprint on which the systems architect can base the following:

- *Prioritization of activities:* Decide which parts of the system will be built first, second, and so on.
- *Total business requirements:* Ensure that each system serves the major purposes of the business and relates to other systems in an efficient manner.
- *Reuse of code, data:* The wheel (analogically—the work of systems developments) must be invented and built once, not many times as is the case when systems are unmodeled and unplanned.

Figure 1.5

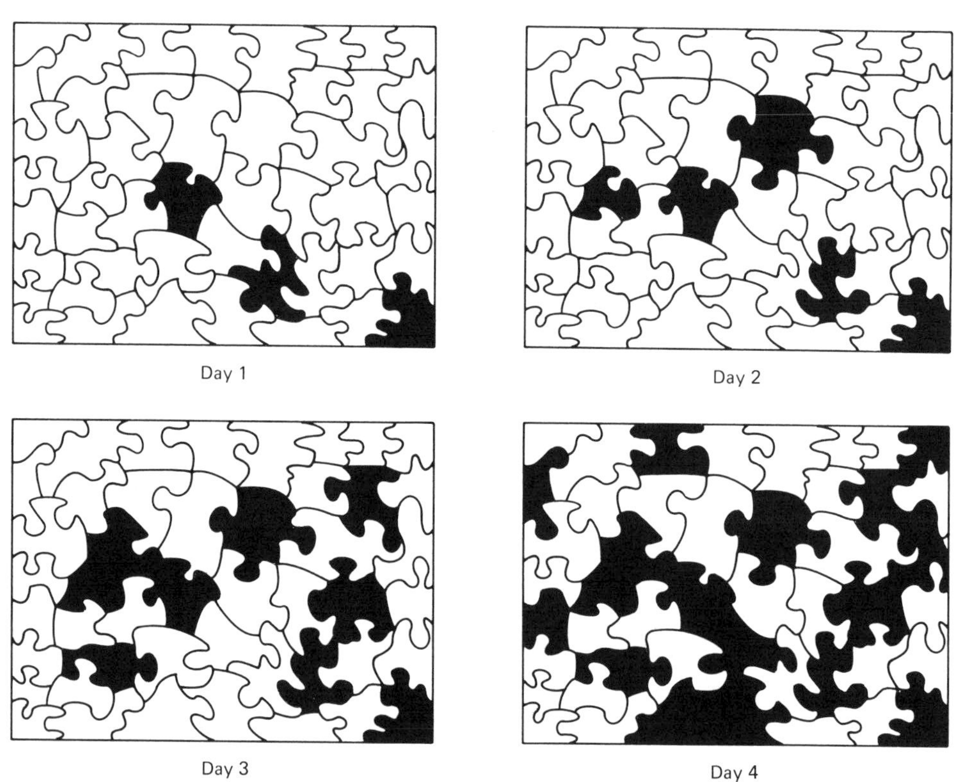

- *Domain:* Provide a definition of domain—a clear-cut definition of who has what responsibilities.

The business-based model is built by liaison between the data processing systems architect and the user. Active participation by both parties is mandatory for the long-term success of the plan. Without the user's participation, the building of the business-based model becomes a technical, data processing activity. Without data processing participation, it is unlikely that the model can be used for a systems blueprint. The participation of more than one user is normal when the system's blueprint is to cover a large scope.

THE MODELING PROCESS: AN OVERVIEW

Once a commitment is made to build a business-based model and the participants are selected, what are the first few basic steps?

- The selection of the *scope of integration,* the boundaries of what is to be integrated and what is not
- A decomposition of the business into the basic operational units
- An identification of the primary business view, including data and processes essential to the running of the enterprise
- A rigorous modeling of the data and processes selected
- A "marriage" of the data and processes at the detailed level

Throughout the modeling process (of both data and processes) there is an emphasis on:

- Simplicity
- Relevancy
- A step-by-step progression where any step is only a slight advance from the previous step, and where the next step is well defined
- Usability and blending of short-, intermediate-, and long-term results

Some of the differences between the unmodeled and modeled environment are illustrated in Fig. 1.6, which shows the unmodeled environment to be complex, created in large (usually unsynchronized) increments, and constantly oriented to short-term solutions. The modeled environment is simple and coordinated, created in a step-by-step fashion where no step is drastically larger than the previous step, and where short-term and long-term goals are blended equally.

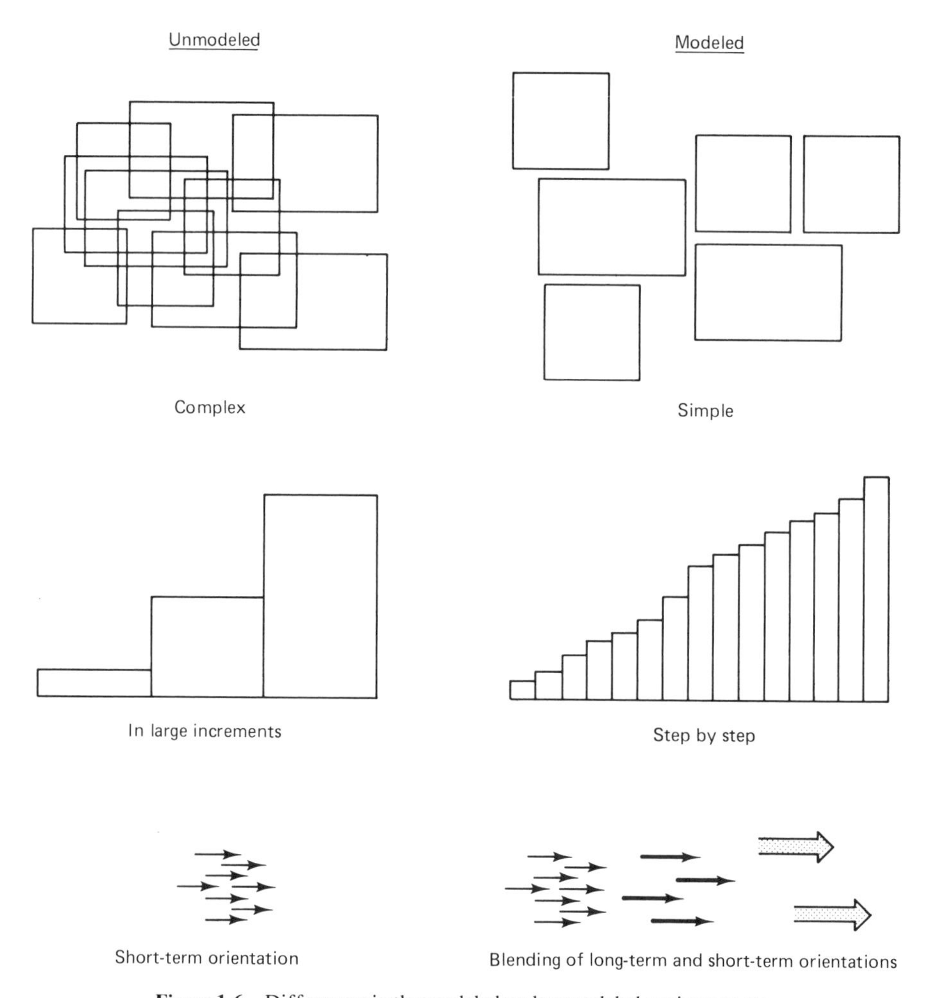

Figure 1.6 Differences in the modeled and unmodeled environments.

DATA MODELING AND ABSTRACTION

Abstraction is one of the underlying concepts at the heart of the effectiveness of business-based modeling. Abstraction refers to the classification of different objects, entities, and processes into common groups based on common characteristics. The importance of abstraction of processes is illustrated by a simple example. Consider the process of withdrawing money from an automated teller machine (ATM) and the withdrawal of money from a teller at a bank window. Although there are basic physical differences between the two processes, at a higher, more abstract level, the processes (i.e., the function accomplished at both places) are the same. They are both the same form of an *account withdrawal.* When the system developer focusses on account withdrawals, however they are done, much unnecessary work can be avoided.

As an example of abstracting data, consider a commercial bank account holder, a grade-school passbook money saver, and a money-market account holder. Although there are basic differences in these entities, at a higher, abstract level, they are all forms of *banking customer*. When the developer focusses on banking customer, rather than the many forms of banking customer, much unnecessary development work is avoided. Abstraction, then, is the conceptual vehicle by which a unified model of a business (its data and its processes) is constructed. Without abstraction, the process of modeling would have to account for many similar, yet slightly different forms of the same thing or process and a unified treatment would be impossible. As a simple example of the usefulness of abstraction, consider Fig. 1.7.

If the data and processes of a system are not abstracted, the systems that are built which use the data and processes are unable to be combined, and much redundant development is the result. For example, in Fig. 1.7 the process of abstraction allows a shop to build a single system—a loan system—whereas an unabstracted approach results in a proliferation of loan systems—FHA home loans, FNMA home loans, signature loans, 30-day loans, and so on. In building a business-based model, it is the process of *abstraction* that allows systems to be built at the highest, most powerful level.

There are some basic techniques of abstraction that are necessary in building a business-based system model. One technique is that of building a chain of abstraction using an "is a type of" connector. Other connectors are "belongs to," "is in the boundary of," and so on. These connectors may be freely mixed within the same chain of abstraction. As an example of a chain of abstraction, refer to Fig. 1.8.

Figure 1.7

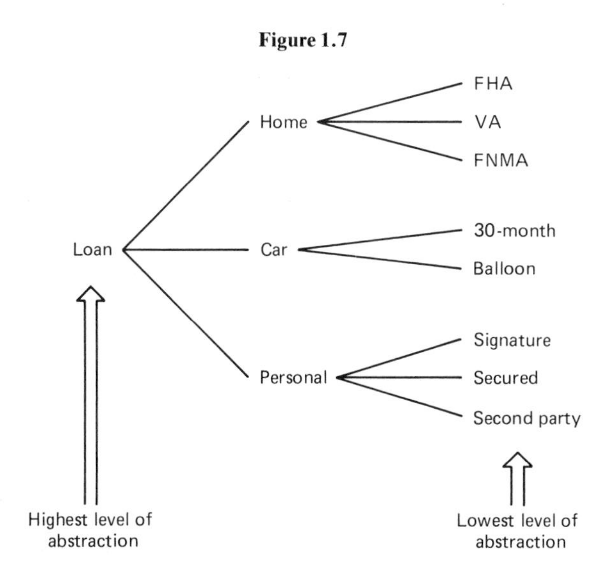

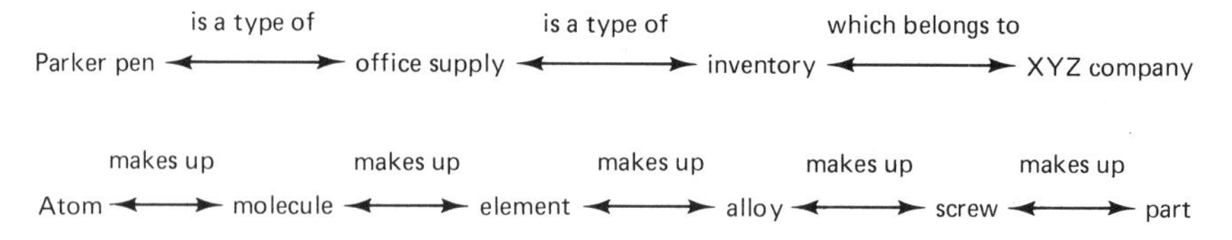

Figure 1.8 Some chains of abstraction.

Even though the *connectors* in a chain of abstraction can be easily mixed, the *types of things being abstracted* cannot be mixed in the chain of abstraction. Specifically, processes and entities should not be mixed in the same chain of abstraction. If processes and entities are intermixed, the resulting chain becomes nonsensical. Figure 1.9 illustrates one such nonsensical mixing of processes and entities. The result of mixing data (entities) and processes in a chain of abstraction is that anything can be abstracted with anything else (thereby losing the usefulness of abstraction).

Another issue relevant to the process of abstraction is how *far* down or up to carry the chain of abstraction. Objects can be abstracted as low as subatomic particles or as high as the universe, but carried to extremes, those levels usually do not make sense. If abstracted too highly, there will be a loss of necessary detail. If abstraction goes too low, unnecessary detail is kept and entails a high overhead (ultimately in

Figure 1.9 Nonsensical chain of abstraction that mixes processes and entities.

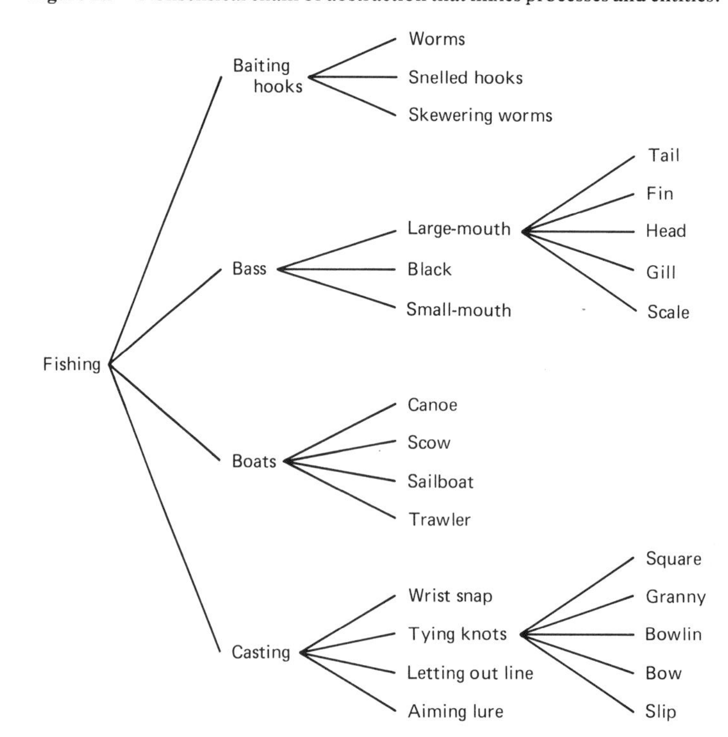

data storage and processing). The proper limits of a chain of abstraction for which a system model is to be built are:

- The upper limit is the level at which all necessary details are kept
- The lower limit is the level at which no unnecessary details are kept

The limits of the level of detail in the chain of abstraction are appropriate to different people at different levels within the business. This difference in perspective is shown in Fig. 1.10. Here the perspective of the secretary is pencils and office supplies. As far as the secretary is concerned, there is no reason to extend the chain of abstraction up to corporate profit. From the IRS perspective, the chain of abstraction includes corporate profits and tax revenues. In only the most farfetched cases will the IRS care about pencils. Thus when building a chain of abstraction, the perspective of the user of the system determines the appropriate limits of the chain of abstraction.

To the uninitiated, the relationship between modeling data and process and abstraction may be obtuse. The importance of abstraction can be illustrated by a simple example. Figure 1.11 shows a collection of data from several systems that have

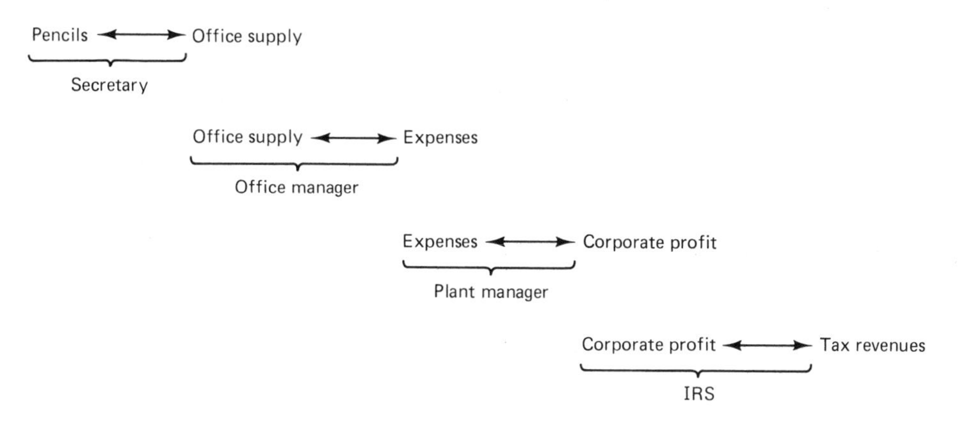

Figure 1.10

Figure 1.11 Unmodeled, unabstracted data from three separate systems.

Passbook account	Christmas club account	Money-market account
Payroll deposit	Withdrawal date	Daily interest rate
Name	Early withdrawal penalty	Minimum deposit
Address	Name	Minimum withdrawal
Minimum balance	Phone	Minimum balance
Current balance	Address	Current balance
Interest (quarterly)	Balance	Name
Interest rate	Carryover balance	Address
Phone	Bonus balance	Account number
SSNO	SSNO	SSNO
Mother's name	Account number	Associated account
Account number		

not been modeled. The process of abstraction is used and the conclusion is drawn that passbook accounts, Christmas club accounts, and money-market accounts are all forms of the same thing—savings account. Abstraction then leads to the consolidation of data, as shown in Fig. 1.12.

Once the data is abstracted and consolidated into a higher form, the system model is more stable because the redundancy of the data definitions is reduced (i.e., the system is built around savings accounts, not passbook accounts, Christmas club accounts, and money-market accounts). Figure 1.12 shows the separation of the unique data and the common data. Only one set of procedures is required to build and maintain the common data as long as the common data is identified. But if the common data has not been identified, separate procedures will be required. The process of abstraction points out the commonality of data and processing.

Figure 1.12

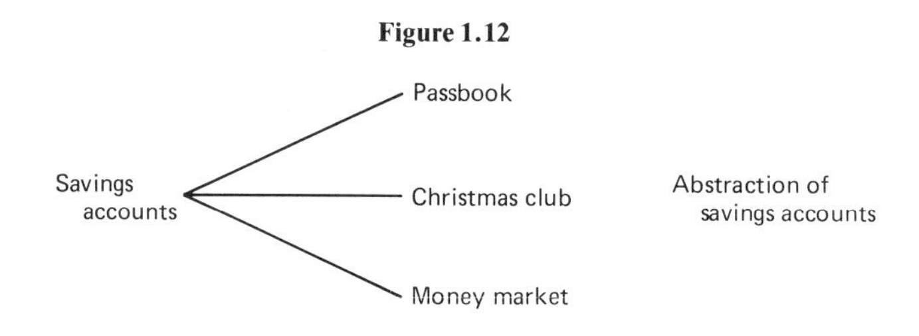

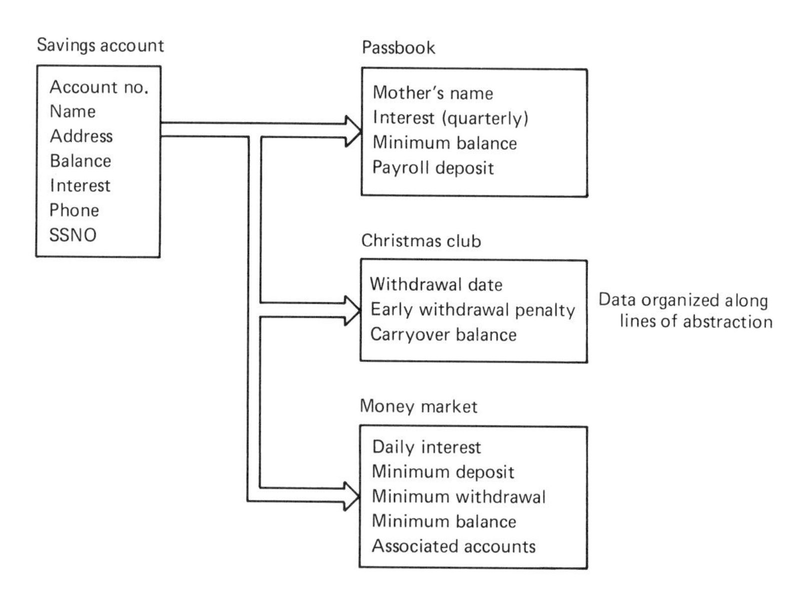

SCOPE OF INTEGRATION

The very first decision that must be made in building a business-based model of a company is the determination of the scope of integration. The scope of integration is merely a statement of what data and processes are to be included in the model and which are to be excluded. This question is *the* most profound question that the system architect will face, because the consequences of an error here are of enormous proportions. If the scope is chosen too small, all processes and entities that need to be included in the model will not be in the model and the parts of the system outside the scope must be developed piecemeal. In essence the wheel must be rebuilt for the data and processes outside the scope. And when the systems outside the scope of integration are built, the major issues of unintegration (redundancy, unnecessary development, maintenance) will surface. When the scope of integration is too large, the design and development effort will be too large, so much so that parts or all of the system being modeled will never be built. In this case, designers are often accused of "blue-skying it."

As an example of the scope of integration being chosen too large, consider a bank. Suppose that the scope were chosen so that *all* banking business were to be within the scope. This would include the normal financial activities (such as loans, deposits, checking, etc.) as well as rental management (of bank property), service bureau processing (such as processing external payrolls), insurance, brokerage, and a whole host of other activities. From a functional point of view there is little similarity between normal banking activities and the collection of rent. So the scope might be chosen to include only normal banking activities. Otherwise, the scope might be too large (depending on the system being built, of course.)

On the other hand, suppose that a system in the banking environment was being built and the scope of integration was chosen too small. Suppose that the scope of the system was only to include loans. The considerations of loan customers as they go to deposit money or open up a checking account would not be taken into consideration. In such a case the scope is probably too small.

Thus the first decision about data modeling is a most important decision. The scope of integration must be chosen just right. What is the "right scope"? It is one that represents:

- A balance between long-term and short-term goals,
- A balance between what can be done and what would be nice to do,
- A balance between the theoretical and the practical.

In most cases the scope of integration is large enough that the system must be built and implemented in phases. It is here that the model serves as a road map to what should be developed first and the relationship between the different parts of the model. This is depicted symbolically in Fig. 1.13.

Figure 1.13 Planning for the implementation of an integrated system.

Typically, a scope of integration represents *only* a single mode of operation. A mode of operation is simply a related collection of processes that operate in a similar manner and operate distinctly differently from processes in other modes of operation. Some examples of modes of operation are the operational mode, the decision-support mode, and the archival mode. Other splits of processing into different modes are the batch and on-line modes. A scope of integration can be defined for the operational mode and another scope of integration for the decision-support mode. Figure 1.14 illustrates the different modes and how they are separated by different scopes of integration.

Note that data will flow from one scope of integration (or mode of operation) to the next. For example, periodically, operational data will be stripped and sent to the decision-support or archival modes. Across the different modes data and processes may well be unmodeled and unabstracted, but within the same mode data and processes will be modeled and abstracted. For example, in the decision-support environment, data will be aggregated and viewed in many different ways. In a bank the vice-president of planning may want to look at the general rise or fall of account balances, the monthly comparison of activities to former years' activities, or the number of loan applications that have been rejected for this year and for previous years.

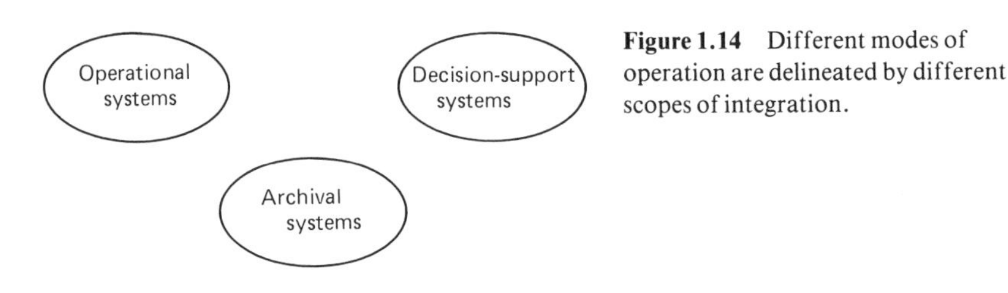

Figure 1.14 Different modes of operation are delineated by different scopes of integration.

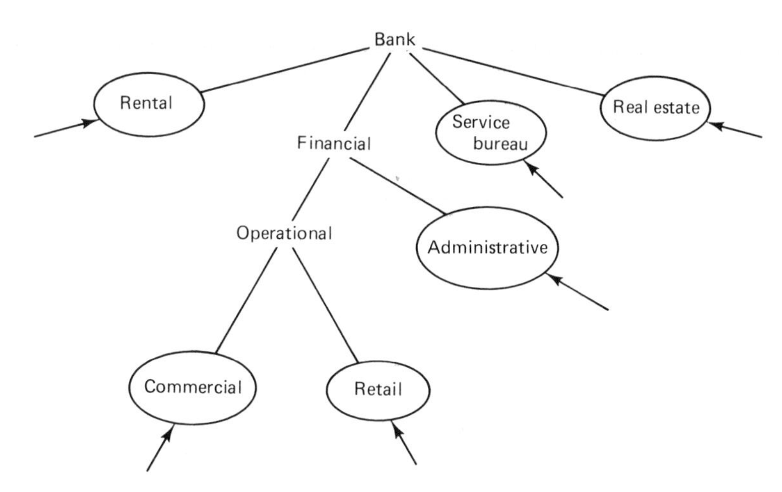

Figure 1.15 Different scopes of integration based on mode of operation or functional division of data.

While these views of data are very much part of the decision-support scope of integration, they most likely are not included in the operational scope of integration. Operational data includes only basic activities such as account activities, interest calculations, and the opening and closing of accounts. To derive the decision-support views of data, raw operational data is passed to the decision-support environment and, once in the decision-support environment, is shaped to meet the decision-support needs.

Another way that the scope of integration is often determined is along broad functional lines. For example, operational systems are normally separated from administrative systems by different scopes of integration. Mixing personnel and benefits systems with bank teller and ATM systems is normally nonsensical. It is therefore normal for system integration to occur in several ways in a large corporation. Consider a large bank that has financial activities, rental activities, service bureau activities, and real estate activities. The different scopes of integration for the bank are shown in Fig. 1.15. Between each scope of integration there can be a flow of data or an aggregation of data at a higher level. As an example of aggregation, in constructing the bank's balance sheet, the expenses and losses within each scope must be calculated, so each business the bank has must report profits and losses. Each scope of integration then passes this information up to a higher level.

SYSTEM MODELING: TERMS AND TOOLS

There are many ways to build a business-based system model, and many tools and techniques. The following represents the set of tools, terms, and techniques that will be used throughout this book and are considered to be most effective.

- *User's environment:* The organization or system as it relates to the business of the enterprise that is being modeled, sometimes called the "real world." The user's environment is where activity occurs that needs to be recorded or measured (i.e., computerized). Typical user's environments are found in banks, manufacturing plants, department stores, telephone companies, government departments, and so on.
- *Entity:* Something about which information needs to be recorded. An entity is unique within the user's environment. No two entities in the same user's environment classify the same object, and everything in the user's environment is classified by an entity. An entity is represented by a circle or oval, as shown in Fig. 1.16.

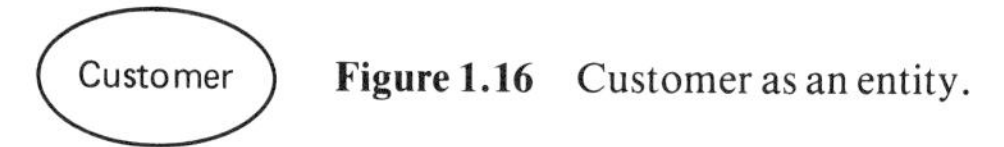

Figure 1.16 Customer as an entity.

- *Relationship:* The association between two entities. It may be 1 to 1, 1 to *N*, *M* to *N*, or nonexistent. It often carries a verbal description. It is described by arrows, which also indicate whether it is 1:*n* or *m*:*n*, as shown in Fig. 1.17. In this figure a customer may have *n* accounts, but no account is for more than a single customer. A part may have *n* suppliers, and a supplier may supply *n* parts.

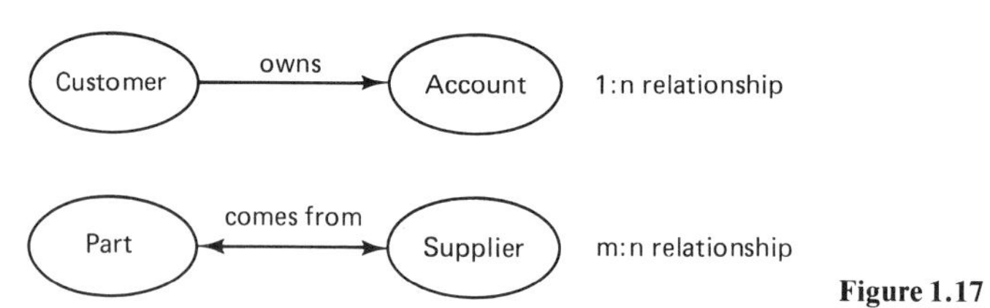

Figure 1.17

- *Entity-relationship diagram (ERD):* A collection of related entities and relationships that belong to the user's environment. Figure 1.18 illustrates a simple ERD.

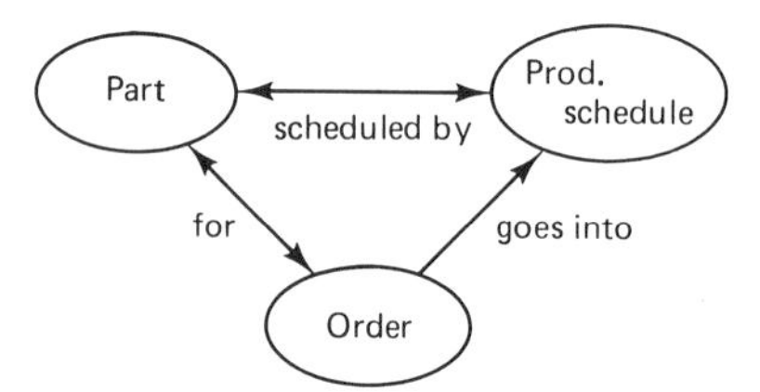

Figure 1.18 An ERD.

- *Element (or Data item):* The lowest unit of information relating directly to the user's environment. An element models one piece of information in the user's environment. An element is normally referred to by the name of what it represents. Figure 1.19 describes some common elements.

Name	Address
Age	Part number
Account	Quantity on hand
.	.
.	.
.	.

Figure 1.19 Some common elements.

- *Data item set (DIS):* A related collection of one or more elements or data items. The contents of the elements or data items in a data item set are determined by the data as it exists in the user's environment. A data item set represents an expansion of an entity. The usual symbol for a data item set is a boxing of elements or data items, as shown in Fig. 1.20.

Name
Address
Phone
SSNO
Age
Sex
Height
Weight

Figure 1.20 Common data item set.

- *Key:* An element or data item that uniquely identifies a data item set. The key may be used for distinctive identification (where the data item owning the key is the only distinct data item set among all occurrences in the data item set) or generic identification (where the data item is one of a class of occurrences in the data item set). The two types of keys are illustrated in Fig. 1.21. SSNO, which stands for social security number, serves to identify any given data item within the data item set distinctly. No other occurrence of a data item set may exist with the same value as that for SSNO. Note that *any* element that is not a nonunique key can be used for a nondistinct key. For example, sex is used as a nondistinct key. Many occurrences of different data item sets may exist with the same value of sex (i.e., either M or F).

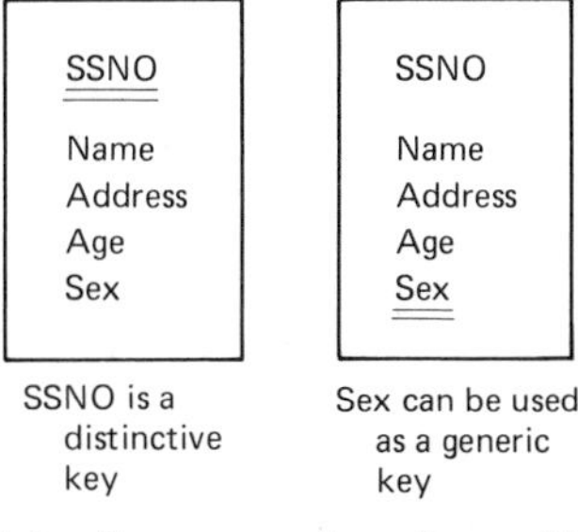

Figure 1.21

- *Attribute:* An element or data item in a data item set that is not a distinctive key. In other words, attributes include all nondistinctive keys in a data item set. Attributes are shown in Fig. 1.22.

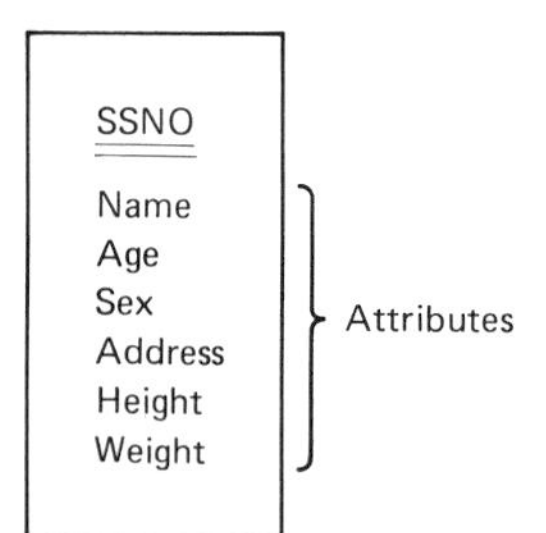

Figure 1.22

- *Characteristic:* Information about a data item set other than that which describes the elements that belong in the data item set. Typical characteristics are the number of occurrences of the data item set, the frequency of occurrence, the cost of unavailability of the data, and so on. Characteristics are documented as shown in Fig. 1.23.

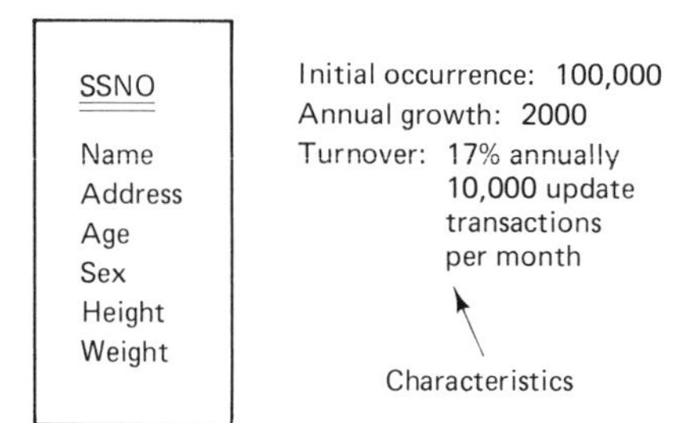

Figure 1.23

- *Connector:* A vehicle for relating two data item sets. For example, the data item set, part number, and supplier are related by one or more connectors. Note that connectors do not imply any particular physical implementation. Connectors can be implemented hierarchically, relationally, by a network, etc. Connectors are distinguished by a single underscore, as shown in Fig. 1.24. Note that the connector does not imply any particular data base management system implementation.

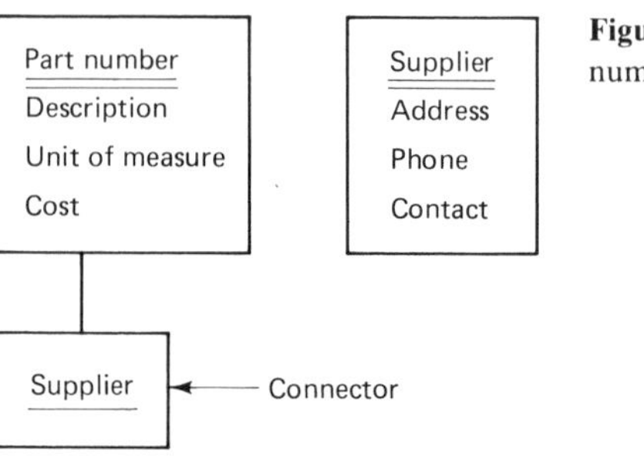

Figure 1.24 Connector from part number to supplier.

- *Recursion:* A data item set or entity that is connected or relates to itself. For example, in the manufacturing environment, an assembly relates to a subassembly, or a cost center relates to its parent cost center through recursion. Recursion of entities is shown by a self-pointing arrow, and recursion of data item sets is shown by a connector pointing to itself, as shown in Fig. 1.25.

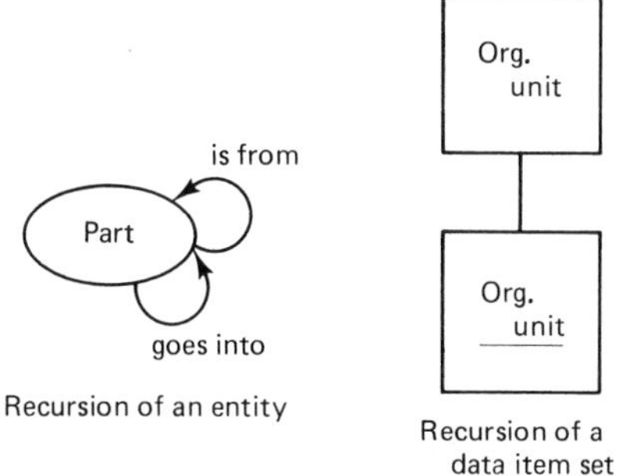

Figure 1.25

- *Scope of integration:* The limitation of the system model. The scope of integration represents the boundary after which the user's environment is not modeled. The scope of integration for a bank would typically include accounts, customers, services, and so on, and would not include such things as political organization information, religious information, racial information, and so on.
- *Dimension:* A single view of data as perceived by someone transacting business in the user's environment. For example, a bank teller's dimension of data would include activity that relates to an account, while an auditor's dimension of data would include activity that has occurred for a branch of the bank for a given period of time. A dimension is modeled by an ERD.
- *Primary dimension* (or *Primary business-based dimension*): The dimension representative of the most basic business activity of all the users within the scope of integration. For example, the primary dimension of a department store includes a customer, a product, a salesperson, and so on. A primary dimension is shown in Fig. 1.26.

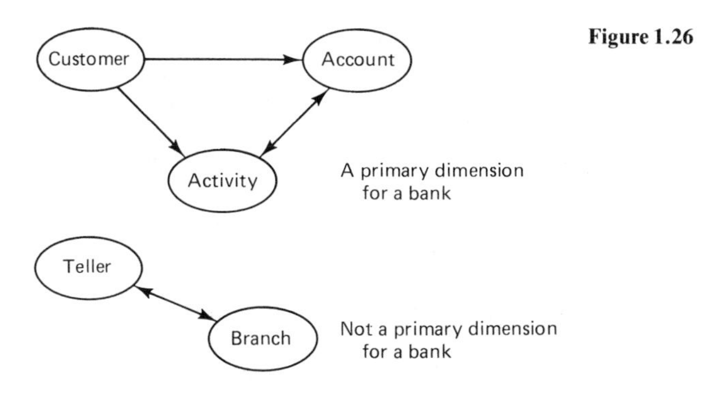

Figure 1.26

- *Entity dimension map:* A diagram of all the dimensions in which a single entity participates. In some cases it is useful to create a data item set dimension map as well as an entity dimension map. An entity dimension map centers around a single entity. For example, in the manufacturing environment, a part belongs to the different dimensions of accounting, production scheduling, inventory, engineering, shop floor control, order processing, and so on. Generally, an entity dimension map is a "working document," and as such may or may not be formalized. An example of what an entity dimension map is and is not is shown in Fig. 1.27.

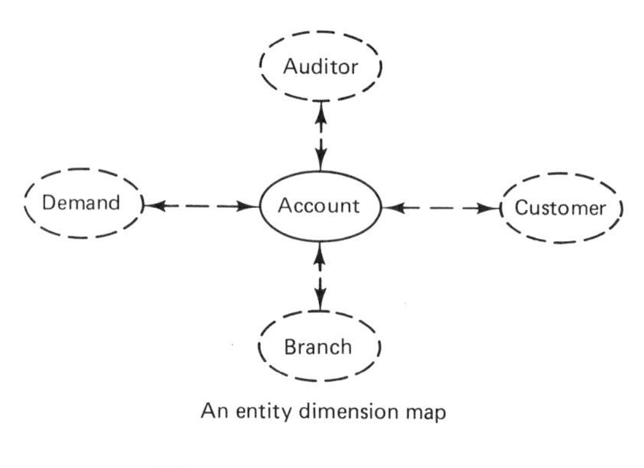

An entity dimension map

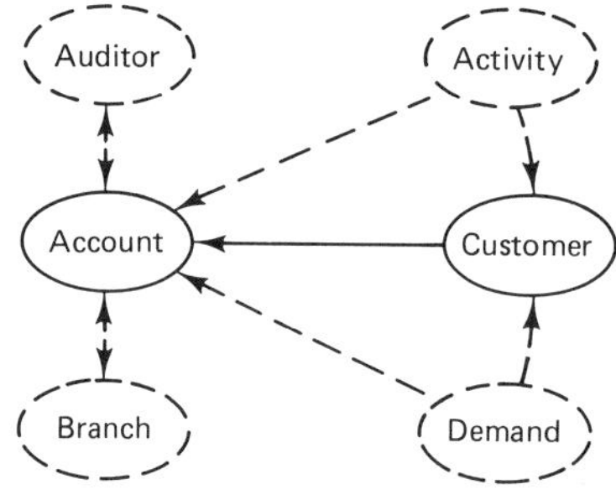

Not an entity dimension map

Figure 1.27

- *Data view (DV):* The same as a dimension, except that DVs are derived in a bottom-up manner, whereas dimensions are derived top-down.
- *Basic processing model (BPM):* An outline of the major functions being accomplished within the scope of integration. All functions that are accomplished within the scope of integration are described by at least one BPM function. A BPM (like all process models) is described in terms of a business cycle. The business cycle is determined by the business being modeled. A cycle contains processes and the flow of activities between the processes. The corresponding level of data design is the ERD. Figure 1.28 illustrates a simple BPM.

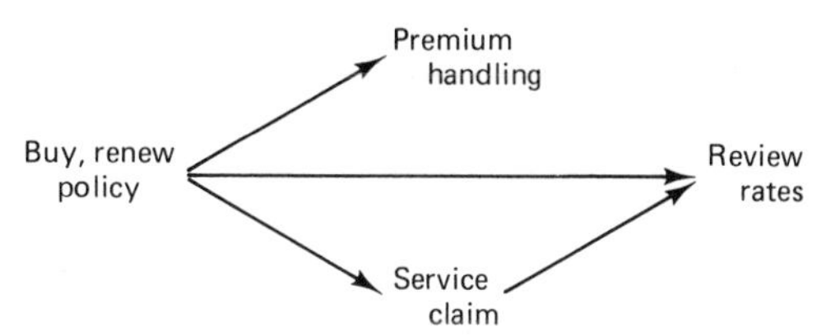

Figure 1.28 Simple basic processing model.

- *Functional processing model (or Functional model) (FM):* An outline of the function achieved by one of the major processes identified in the basic processing model. The functional processing model expands each of the activities or processes outlined in the BPM. The corresponding level of the data model is a data item set. For an example of a functional processing model, refer to Fig. 1.29.

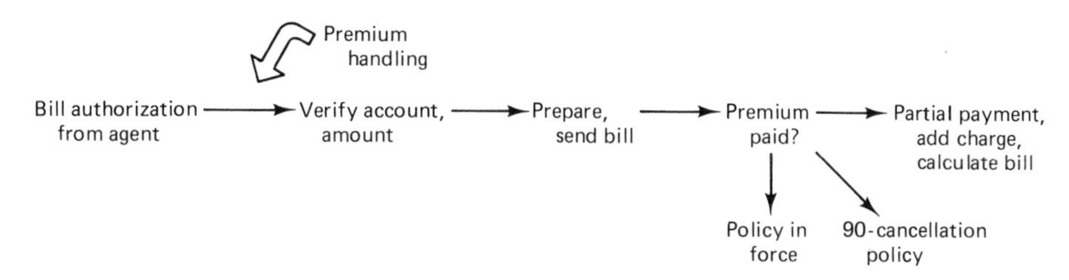

Figure 1.29

- *Detailed processing model (DPM):* An outline of the processes that must be accomplished to satisfy each step in the functional processing model. The detailed processing model goes down to the level of detail at which programming specifications can be drawn up. The corresponding data model is a physical model of the system being implemented.

OTHER TOOLS AND REPRESENTATIONS

A business cycle represents the execution of the normal functions that occur within the scope of integration from the initiation of the first activity until completion.

- *The © and @c symbols:* Indicate end of cycle ©, start of cycle, and drop out of cycle @c. These symbols help show the flow of the business cycle, as shown in Fig. 1.30.
- *FOR loops:* Indicate iterative processing; refer to Fig. 1.31.

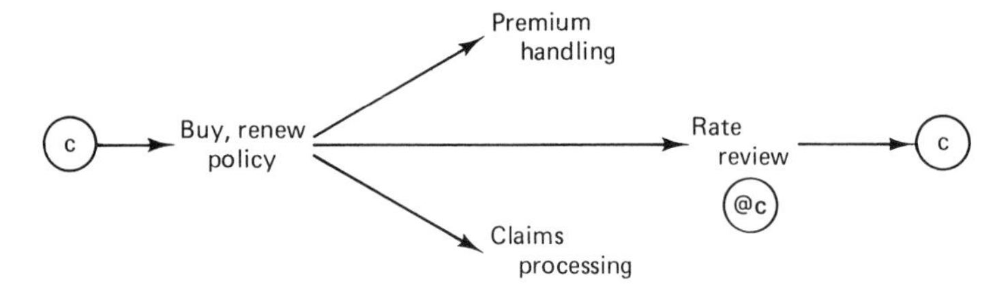

End of cycle, beginning of cycle

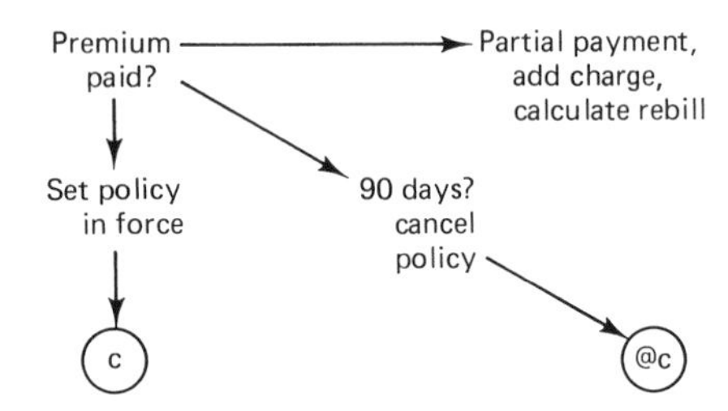

End of cycle and dropout of cycle

Figure 1.30

```
Order process     For line 1 to n:
                    Verify part number
                    Locate part
                    Allocate part
                    Assign to staging area
                    If not enough parts:
                      Set up back order
```

Figure 1.31

- XXXXXXXXXXX: A process is used here but is defined elsewhere. This type of process referral is for subroutines and low-level detailed processing that are common across the user's environment. Figure 1.32 illustrates the use of process referencing.

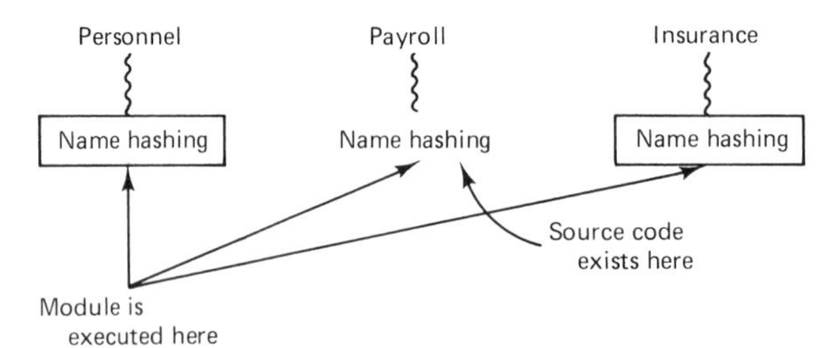

Figure 1.32

- *"Is a type of" relationship:* Indicates that a data item set has a further logical division. When "is a type of" is specified, all types of entity or DIS must be categorized. This relationship designation corresponds directly to the logical connector of the same name that is used in the process of abstraction. Figure 1.33 illustrates an example of the "is a type of" relationship. In the figure there are four types of accounts—loans, savings, DDA, and trust—and there are three types of loans—car, home, and signature.

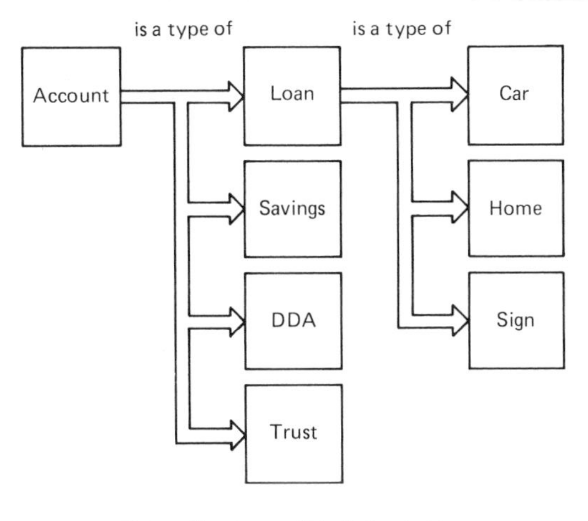

Figure 1.33

- *Aggregation/subset:* An aggregate/subset of an entity is denoted by a dotted line separating the various parts of the entity. The notation means the entity can be viewed individually or in aggregation. Although infrequently used, this notation comes in handy on occasion. Refer to Fig. 1.34 for an example.

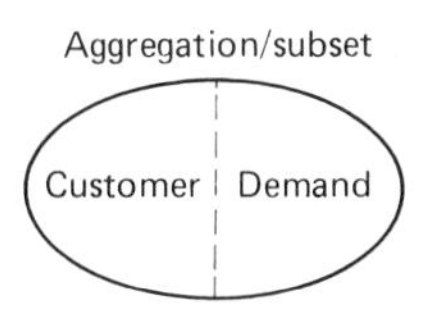

Figure 1.34 A customer forms the basic unit of demand; demand is formed by a collection of many customers.

- *Other dimension relationships:* For entities or data item sets, a relationship to another dimension is denoted by dotted lines. Figure 1.35 illustrates a multidimension relationship.

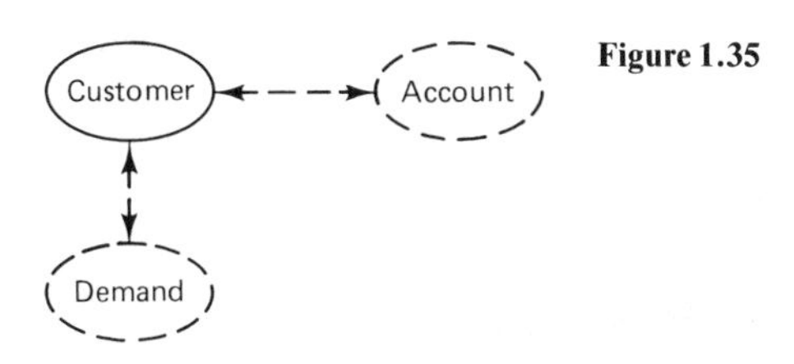

Figure 1.35

- *Global scope of integration map:* A diagram of the different scopes of integration that belong to a business. In the case where a single scope of integration encompasses all the business, there is no need for this type of map. The map is used primarily for identifying interfaces between different scopes.
- *Global ERD map:* A diagram of the different ERDs that make up a single scope of integration. In the case where there is one ERD in the scope, the map is not needed. The map is most useful for identifying common entities and processes and interfaces between different dimensions.
- *Consolidated process model:* The resulting process model produced by merging the common functions found in the DPMs that represent different dimensions.

SOME BASIC MODELING TECHNIQUES

What happens when data is modeled such that it depends on two or more units (entities, DISs, etc.) of data for its existence? (In some circles, this is called "intersection" or "junction" data.) To better illustrate this case, consider the data in Fig. 1.36. In this case, a supplier handles orders, and when a part is needed, an order is created. In either case, the existence of an order depends on *both* a part and a supplier. This case is best handled by creating order as its own entity, as shown in Fig. 1.37. The advantage of this form of the data is that the relationships are maintained and the data—order—is consolidated into a single form. Does this imply that similar data cannot legitimately exist separately? Not in every case. Consider the example shown in Fig. 1.38. In this figure data relating to limits paid by the insurance company are shown as existing separately in two places. But a closer look shows that the limits apply to insurance and claims in a very different way, and the data does not truly intersect. Probably, the name "limits" should be refined in one or both cases. Another useful technique is the connector to describe data belonging to a different dimension.

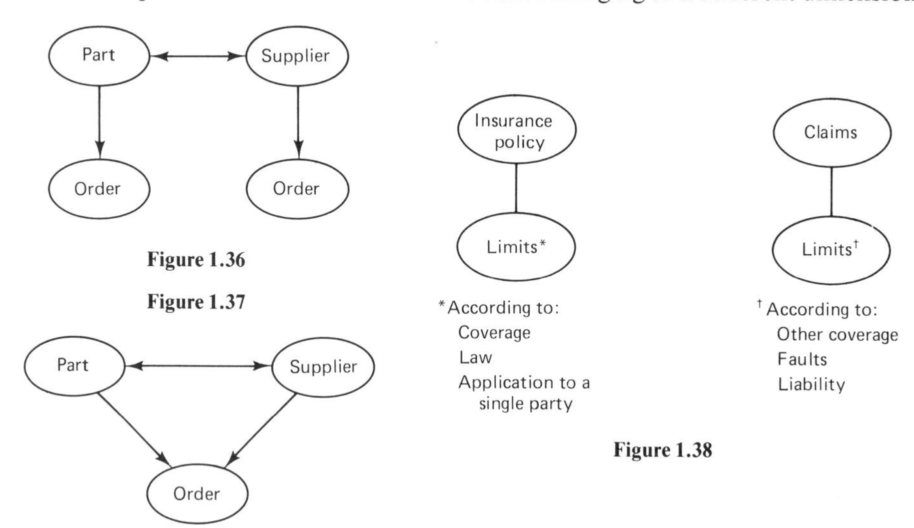

Figure 1.36

Figure 1.37

Figure 1.38

It has been seen that data can be connected across dimensions prior to consolidation. Another use of connectors is in the crossing of modes of operation by use of an underscored connector, as shown in Fig. 1.39.

Another standard practice is occasionally to show multiple occurrences, usually of a data item set. The technique is to show a "shadow" behind the first occurrence of the data item set, as shown in Fig. 1.40. This shows that multiple activity dates exist for an account.

Another modeling guideline is that data at the ERD or DIS level, or processes at the BPM, functional, or DPM level are abstracted to the highest level. This means that, within an ERD, for example, no two entities "are types of" the same thing. For example, consider the three entity relationships shown in Fig. 1.41. In the first case, there are three entities: a parent, a son, and a daughter. But a son and a daughter are a type of child, so the second representation is better than the first (because it represents a higher level of abstraction). In the second representation, a parent and a child are both types of a person which is encompassed by the third case (which is at the highest level of abstraction). The chain of abstraction can be depicted as shown in Fig. 1.42. Since a person that has parentage relationships represents the highest level of abstraction, the third representation is the preferred choice.

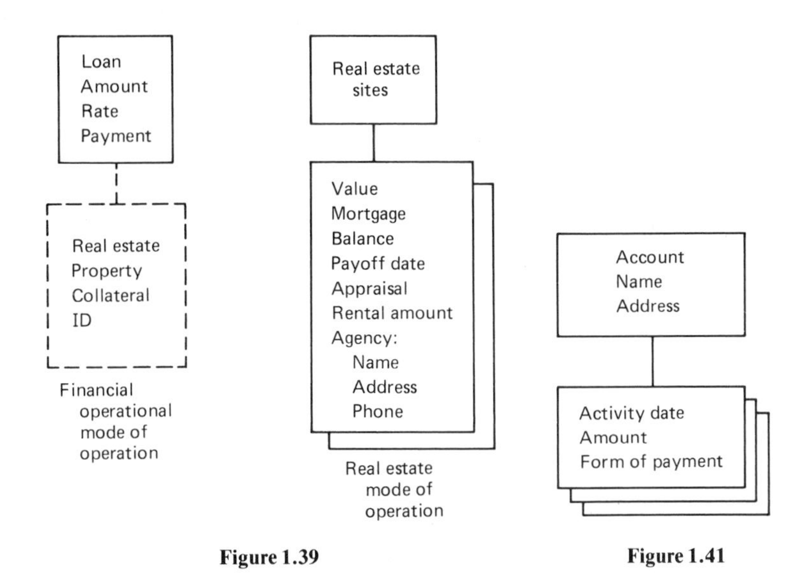

Figure 1.39

Figure 1.41

Figure. 1.40 In this case activity can occur multiple times for the same account.

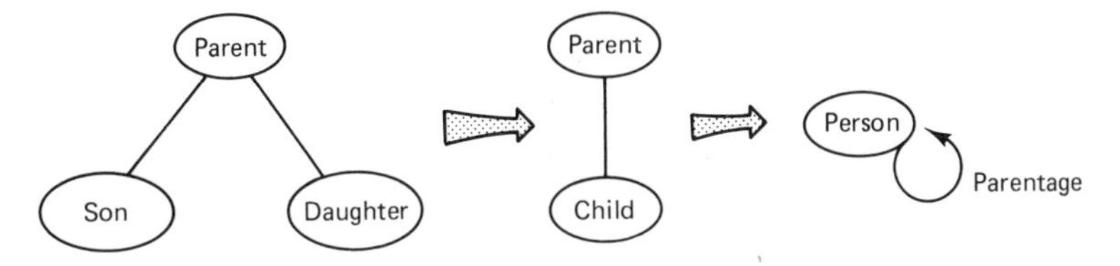

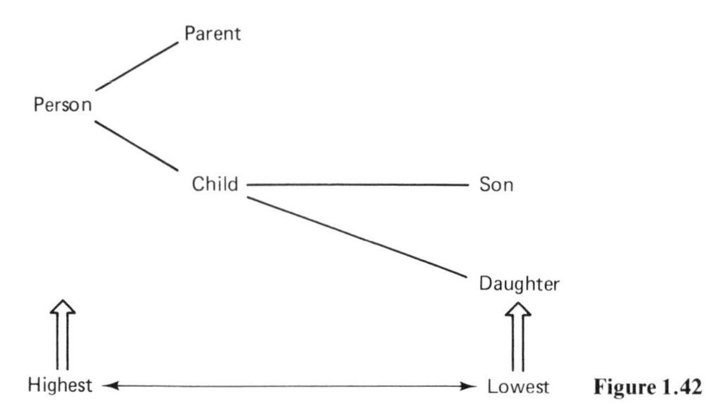

Figure 1.42

MODEL FLOW

The general flow of the modeling process for building business-based systems is depicted in Fig. 1.43. The flow shows that there is a single set of activities from the identification of the scope of integration to the point of entity and DPM identification. At that point there are dual data and process modeling activities, and there are parallels at each step between data-modeling activities and process-modeling activities. At the end the models are "married," verifying the quality and completeness of the business-based model produced.

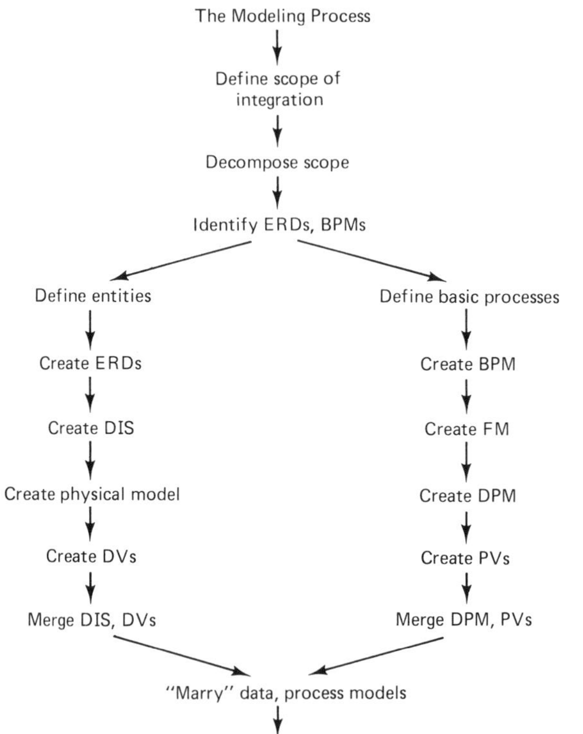

Figure 1.43 The modeling process.

Figure 1.43 is somewhat misleading in that it is normal to reiterate all or part of the modeling process, and the figure does not show that reiteration. For example, the discovery of an error at the process view (PV) level may actually cause the scope of integration to be reconsidered. Hopefully, not *all* of the scope must be reconsidered. (If, in fact, a problem at a lower level of modeling does cause the entire model or major portions of the model to be dismantled, it is a sign that there is something fundamentally wrong with the modeling process.)

Although Fig. 1.43 shows the generic flow of an integrated design, it does not show how all of the various modeling activities interact. The integrated design of a system on a step-by-step basis is better depicted by Fig. 1.44, which shows the information systems architecture (ISA)—that is, each step of the integration and how the steps fit together. The ISA checklist at the end of this chapter further describes the interaction.

The first activity shown in the Fig. 1.44 is the creation of a map of the global scope of integration. This activity may be optional if the system to be integrated is small or if the map has already been built. The next activity is the definition of the scope of integration. If there are major functional differences in the business being modeled or if there are different modes of operation to be modeled, there may be multiple scopes to be defined here. This activity is not optional.

After the scope(s) is defined it is decomposed, usually along the lines of the organization chart. After the decomposition is done, at each lowest point in the decomposition the data and processes that are common are identified and are selected for the primary business-based ERD and BPM. The decomposition and synthesization modeling activities are not optional.

The next step is the creation of a global ERD map. This step is optional, as some scopes of integration contain only one ERD. At this point the activities of data and process modeling take different paths but are done in parallel.

The first ERD to be developed is the primary business-based ERD, as identified after the synthesis of the decomposition. Other ERDs can be developed in parallel or at a later point in time. If the ERDs are developed later, the modeling process cannot proceed past the point of consolidation because all ERDs contribute to the consolidation process. Whether the ERD is the primary business-based ERD or not, the process of modeling is the same. The first activity of ERD modeling is to identify the entities that belong in the ERD. Then the entities are defined. After definition the relationship between entities is identified. These three activities are normally done all at once and are frequently reiterated. The result of this phase of modeling is a completed ERD.

Each entity in the ERD is then expanded into a DIS. After the DIS is created, if there is a commonality with DISs from other ERDs, a consolidated DIS is created. Otherwise, the merger with the bottom-up DVs is ready to begin. DVs are created strategically (i.e., they are created selectively and representatively) and are nothing more than the bottom-up view of data. The detail of the DVs is used to verify the completeness of the detail of the DIS. The result is a physical modeling of data, ready for translation into a physical data base design.

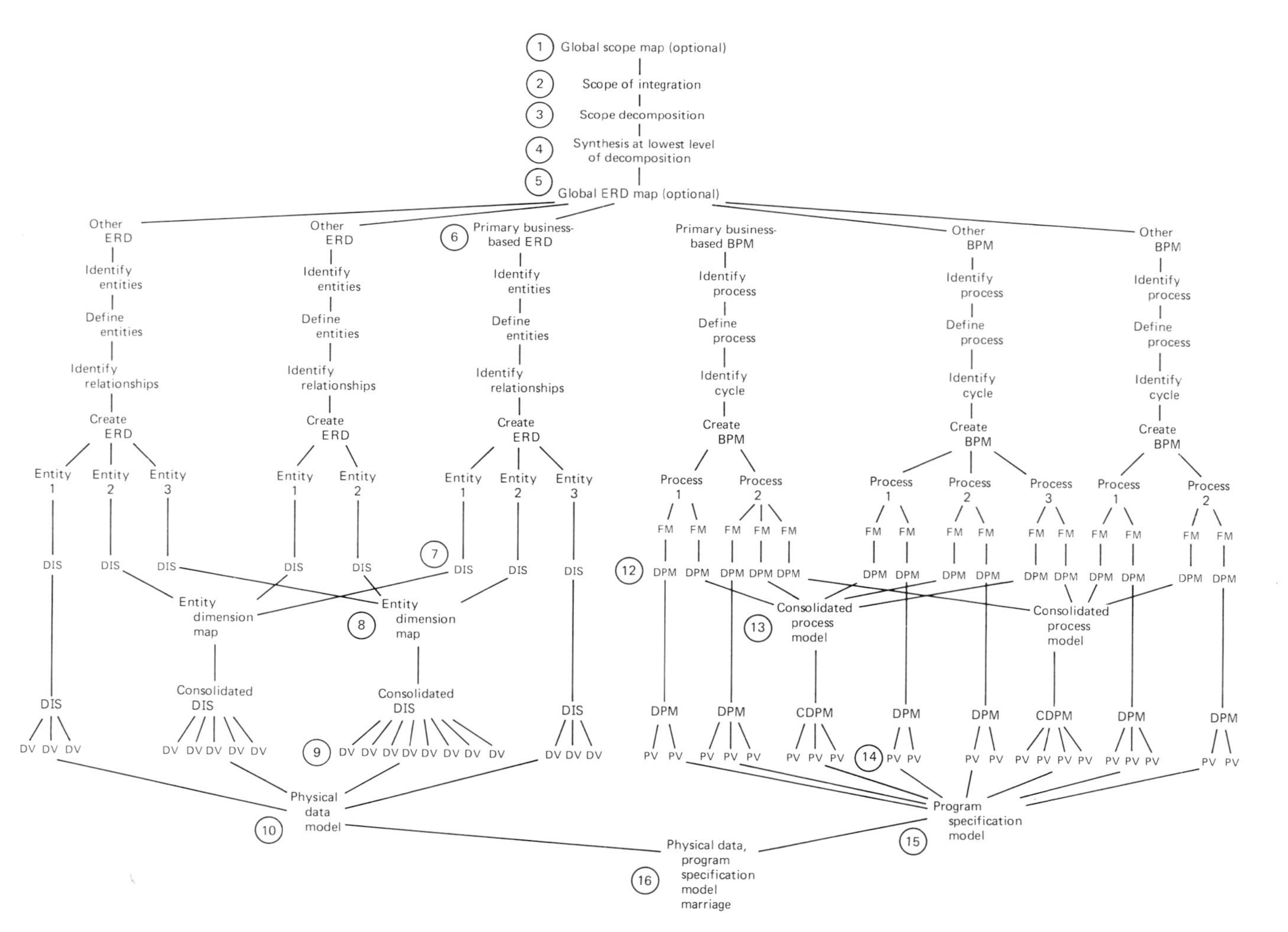

Figure 1.44 The information systems architecture.

In parallel with the ERD creation is the BPM creation. The primary business-based BPM is the first process to be modeled. Processes representing other dimensions can be modeled either in parallel or at a later time. If modeled later, the consolidation process cannot proceed until all dimensions are modeled. In any case the modeling process for a BPM is the same whether the primary business-based process or some other process is being modeled or not.

The BPM modeling process begins with an identification of the major processes of the BPM. After identification the processes are defined and the business cycle in which they participate is identified. These three activities are done in tandem and are reiterated frequently.

Each major process defined in the BPM is then broken into an FM. The FM is merely an expanded description of the process. From the FM the detailed process model is built. Just as the FM expands the processes of the BPM, the DPM expands the processes of the FM.

After the DPMs are built, a consolidated process model is built for common processes. Some DPMs participate in a consolidated process model and some do not. Once the DPMs and consolidated DPMs are built, they are ready to be compared to the bottom-up-constructed PVs. PVs (like DVs) are built strategically and are used to ensure that the DPM accomplishes all the functions that need to be accomplished.

The result of the PV/DPM merger is the program specification model. From this model program specifications can be built. After both the physical data model and the program specification models are built, they are analyzed together. The program specification model uses the data model to determine whether the data needed for execution in fact will be present. The physical model uses the program specification model to determine whether the data present is created, updated, accessed, and deleted.

The business-based system modeling process just described is essentially a top-down process with bottom-up verification. The emphasis on top-down form is dictated by the needs of integration, which must first consider the broadest perspectives of the system being built.

IN SUMMARY

In this chapter the objectives, directions, and philosophies of the information engineering required in the building of the business-based model have been discussed at an overview level. The stability of systems is the primary goal of information engineering. Stability is achieved when the systems built are based on a cohesive model of the data and processes of the business. The business-based model in and of itself has limited use. *Why* to build the model and *what to do* with the model after it has been built are at least as important as the building of the model itself.

Effective business-based modeling requires management understanding and commitment. In addition, user participation is vital. Without user participation business-based modeling becomes a backroom technical data processing exercise. In addition, organizational attitudes must be changed from the traditional local perspective to a global perspective. The resulting business-based model of data and processes must exhibit *at least* the following:

- Simplicity
- Relevancy
- Blending of short-, intermediate-, and long-term goals

An underlying concept vital to a sucessful model is that of abstraction, where data and processes are grouped at the highest level. Another important concept is that of the scope of integration, within which the model's data and processes lie. Finally, many terms and tools were described that will be used in building the business-based system model.

PROJECT STUDY

The Anaconda Banking Company (ABC) has been in business since 1872, when timber, mining, and farming interests began to solidify in the Rocky Mountain West. Based in Denver, ABC has branched out throughout Colorado, and recently has affiliated with other western banks in Wyoming, New Mexico, Utah, Arizona, Nevada, Idaho, Washington, Oregon, and California.

ABC provides a combination of commercial banking services and personal banking services. The financial services that ABC provides are divided evenly between commercial and personal. In addition, ABC has diversified and offers various nonfinancial services such as payroll processing, independent data processing services, accounting, cash flow analysis, and other miscellaneous services.

Throughout the years ABC has acquired property (primarily real estate) by several means. One way has been through building new facilities for normal banking operations. ABC has large offices in Denver, Salt Lake City, Cheyenne, and Spokane, as well as many smaller offices throughout the West. Most of the banking operation facilities are used by ABC, but the location of some offices are so choice that ABC is better off leasing the prime office space to private renters and moving their own operations personnel to less desirable locations.

The second way that ABC has acquired property is through a variety of channels, such as moving personnel and guaranteeing the sale of real estate, foreclosures, equity trades, and so on. Over the years ABC has acquired a massive amount of property which it manages. There are some smaller businesses owned by ABC, such as their leased-line business. In building a network between Denver and the outlying offices, ABC has constructed a large network. The primary usage of the network is for the overnight flow of data. But during the day the network sits idle, so ABC allows other businesses to transport data during daytime hours. As an example of another small business that has arisen, ABC has hired several contract programmers to write software for them. Much of the software is of a general-purpose nature, such as report generators and a data base screen definition program. ABC allows the contract programmers to lease the software to other firms for a royalty.

ABC has been into data processing for many years. In the early 1960s ABC used card equipment: sorters, card punches, wired boards, and so on. In the 1970s IBM's general-purpose hardware and software was purchased, together with data base software. ABC built most of their systems in house, a system at a time. Among the many systems that ABC has are mining loan and depreciation, hourly payroll processing, Wyoming payroll processing, Utah payroll and benefit processing, branch savings and deposit, home loans—Arizona, car loans—Utah, Colorado payroll processing, employee benefits, car loans—Oregon, payroll deposit, farm loans—federal guarantee, commercial loan collection—California, car loans—Wyoming, salaried payroll and benefits, private IRA, commercial cash management, interstate bankcard, correspondent bank transfers, car loans—New Mexico, home office cash management, commercial DDA, private DDA, interstate DDA, mining reinvestment, farm land bank, interstate convenience banking centers, branch teller systems, ATMs, up-scale banking services, and government bond investment.

In the early days of data processing ABC seemed to be content with the data processing department. Although systems were slow in being developed, at least they came up eventually. But as more and more systems were built the user satisfaction level began to drop. More time was taken to build systems, more time was taken in fitting new systems with old systems, and the ratio of maintenance to development climbed dramatically. Things have reached the point where development is at a snail's pace and yet there are many more data processing personnel than there ever were before. In fact, the demand for new programmer/analysts climbs inversely with the users' satisfaction with the systems being produced.

Recently, ABC has considered a change in account number from 12 digits to 13, and all applications groups that would be affected were invited to discuss the impact of such a change. There were so many people in attendance that the meeting had to be held in an auditorium.

ABC had hoped that by going to a data base system, many of their problems would be solved. In fact, some problems were solved. But even though the standard hardware and software that are found throughout the industry are used, data base has proved to be more of a headache than anything else. The data processing budget has been climbing out of sight annually. Even though ABC hires many programmer/analysts each year, the hardware budget is increasing even faster. In 1959 data processing represented less than 0.01% of the operating budget of ABC. By the mid-1960s data processing still represented less than 0.5% of the operating budget. By the 1970s data processing represented 2.75% of the operating budget, and by 1980 it represented 6.5% of the operating budget.

This seemingly relentless trend (that continuously swallows up more and more resources) greatly disturbs top management because ABC is being squeezed for profits in a day of banking deregulation. As long as data processing represented a small portion of the budget, top management really did not care about its operating efficiency. Data processing was considered to be a technical realm, somewhat tangential to the business of the bank. But in recent years data processing has moved much closer to the business of the bank, especially as the bank tries to compete with other banks and financial institutions. Still the attitude toward data processing is that it is a technical function.

1. Describe the problems faced by ABC in terms of data processing and the business of ABC.
2. Is there a single solution to the problems of ABC?
3. Is the percentage of operating budget spent on data processing normal? What is normal? Can the climb of the budget be expected to level off? Keep climbing? What does this mean to ABC?
4. Is the ratio of maintenance to new development normal? What is normal?
5. Has data base been a failure? What has failed?
6. Is the progression of events described common? Do all shops go through this progression?
7. Is the problem a lack of management leadership? Technical leadership? Of attitudes?
8. Outline 10 steps that need to be taken for ABC to begin to control data processing and begin to cope with its problems.

EXERCISES

1. (a) Describe how models outside data processing are used—in engineering, in marketing, in city planning, and so on.
   (b) When are models useful, in general? When are they not useful?
   (c) When does the expense of model building become too much?
   (d) When a model is used for a blueprint, how should it be kept up to date?
   (e) How can an organization ensure that the model is adhered to?
   (f) When should variances be allowed?
   (g) What happens if an inadequate model is built? What are the characteristics of a good model? Of a bad model?

2. (a) Once the system model is built, how should it be applied to the organization?
   (b) Does the existence of a model necessarily ensure that systems will be integrated? Does the existence and enforcement of a model necessarily ensure that systems will be integrated?
   (c) What conditions must be met to achieve integration?
   (d) Is integration worth achieving? Why? Why not? What is at stake in achieving or not achieving integration?

3. Under each of the following conditions:

   (1) When modeling is done
   (2) When modeling is not done

(3) When systems are integrated
how will each of the following people be affected?

(a) The user
(b) The data administrator
(c) The data base administrator
(d) The user management
(e) The development programmer
(f) The maintenance programmer
(g) The computer operator
(h) The vice-president in charge of finance
(i) The hardware vendor
(j) The software vendor

4. (a) What is the "local" approach? Why is data processing traditionally organized around the local perspective?
(b) When should data processing reorganize around the "global" perspective?
(c) Which perspective should override the other? Should they be balanced? Why? Why not?

5. (a) What happens when system design is done by focusing almost entirely on process design? On data design?
(b) Why is process design applicable to an exclusive focus on processes?
(c) Why is the on-line integrated environment focused on a balance between data and processes? Is it possible to build an on-line integrated environment and focus exclusively on either data or processes?

6. (a) What is the scope of integration?
(b) What is its importance to the integrated environment?

7. (a) Why is abstraction important to the integrated environment?
(b) What is the chain of abstraction?
(c) Create a chain of abstraction for the following two levels up and down the chain:

(1) Data
- A car
- A person
- A planet
- A political party
- A religion
- A pencil

(2) Process
- A manufacturing assembly
- The election of the U.S. President
- The royal succession of England

- Cashing a check
- Catching a fish
- Writing a book
- Adding two numbers together

(d) Why should data and processes never be mixed in a chain of abstraction?

(e) How far, up or down, should a chain of abstraction be carried?

8. To achieve integration, long-term goals must be translated into a series of short-term objectives.

(a) Give an example of two long-term goals that are to be achieved by integration, together with five or six short-term objectives that would support those goals.

(b) How can progress be measured? Is there such a thing as short-term objectives that only appear to satisfy long-term goals? It is possible to create short-term objectives without fulfilling long-term goals? If so, how?

9. (a) Define the following terms.

(1) Entity
(2) Relationship
(3) Element
(4) Data item
(5) ERD
(6) DIS
(7) Key
(8) Attribute
(9) Characteristic
(10) Connector
(11) Recursion
(12) Scope of integration
(13) Mode of operation
(14) Dimension
(15) Primary dimension
(16) Entity dimension map
(17) DV
(16) BPM
(19) "Is a type of"
(20) Functional process model
(21) DPM
(22) Aggregation/subset
(23) Global scope of integration map
(24) Global ERD map

(b) Give two examples of each term.

(c) Discuss the relationship of each term to two or three other terms.

THE ISA PHASE CHECKLIST

1. *Identification of business components.*
   a. Have all components of the business been identified?
   b. If there is only one business component, there is no need to create a global scope map. Go to step 2.
   c. Identify each business component. Show how all the components relate to the organization. Show how each component relates to each other component.
   d. Will each business component be modeled? If so, how will the relationship between components be defined? If there are components that will not be modeled, and if there are relationships between unmodeled components and modeled components, how will that relationship be defined?

   OUTPUT: The output of this step is an identification of the major lines of business that are to be modeled. If more than one major line of business is to be modeled, the relationship between the lines must be identified. The output is short and in plain English.

2. *Definition of the scope. Note:* This activity must be done for each scope of integration.
   a. For the scope of integration to be modeled, define the scope. What lies within the scope? What lies outside it? Is there anything that lies within the scope and another scope? If so, identify the overlap. How will the overlap be managed?
   b. What happens if something that is not defined within the scope in fact should be there? How easy will it be to readjust the scope?
   c. What future (major) changes are imminent? Are possible? What impact will these changes have on the scope? Should they be included? What effort will be required to include them later?
   d. Define all the different dimensions within the scope of integration. Which are most important? Least important? Which dimensions are so similar to others that they should be combined? Which are clearly distinct?
   e. Give a brief description of the primary business-based dimensions.

   OUTPUT: The output of this step is a description of the scope of integration. It should be written in language understandable to management, the users, and data processing personnel. It should be about two or three pages long, certainly no more than five pages.

THE ISA PHASE CHECKLIST (cont.)

3. *Scope decomposition*

   a. Using the organization chart, start with the highest organizational unit and expand the chart. The general line of control is most important. Do not make the decomposition complex by including matrix management responsibilities unless matrix responsibilities make up a significant part of the organization. If matrix responsibilities do make up major parts of the chart, create another decomposition chart.

   b. Decompose the organization to the lowest functional unit. The lowest functional unit is the one where activities beneath it are indistinct. For example, an organizational unit that merely collects money for a bank is not distinct, because the money collected could be for loans, savings, bankcards, and so on. But an organizational unit servicing loans is distinct from an organizational unit servicing checking accounts.

   c. Is there any function in the scope of integration that is not represented in the decomposition?

   d. Is there any function in the scope of integration that is represented in more than one place?

   e. Is there any decomposed activity that is not functionally unique? If so, why has it not been consolidated to the lowest functional unit?

   f. What data is common to most lowest functional units? What processes are common? Is any data/process common to *all* lowest functional units? Is there little commonality at each lowest functional unit? (If more than one mode of operation is within the scope, this may be a real possibility.)

   OUTPUT: The output of this phase is a description of the business to be modeled, broken down to the lowest functional level. The format may be in symbolic form (with lines connecting functions as they are decomposed), in indented outline form, or both. The output may be as formal or as informal as desired. Completeness of decomposition is more important than organization, although the organization should combine as many like functions as possible.

4. *Synthesis decomposition at the lowest level*

   a. Once the scope is decomposed to the lowest functional level, identify data and activities that are common to most lowest functional levels. The data and processes that have been identified form the basis for the ERD and BPM activity.

THE ISA PHASE CHECKLIST (cont.)

b. The data and processes, once identified, are abstracted to the highest level appropriate to the scope of integration.

c. What are the various dimensions that must be modeled? Are all users' perspectives represented by one or more dimension?

OUTPUT: The output of this step is a definition of which data and processes are to be modeled and what dimensions will be used to view the data. The data and processes should be abstracted to the highest level that is appropriate to the scope of integration.

5. *Global ERD map.* (A global ERD map is necessary if there is more than one dimension within the ERD. If there is only one dimension in the ERD, as is common, no map is required.)

   a. Outline all the dimensions. Give each one a name.

   b. If two or more dimensions are so similar that they should be combined, combine them.

   c. Show how each dimension relates to the scope. Show how each dimension relates to each other. (Note that some dimensions may not relate at all to other dimensions.)

   d. Do any dimensions exist that are within the scope of integration but are not represented as a dimension? Why?

OUTPUT: The output of this step is an identification of the different dimensions that are within the scope of integration and an outline of how they relate. If there are many dimensions (more than 15 or so), there is probably a need to consolidate similar dimensions. The output should be simple and straightforward.

At this point *both* data and process modeling are ready to proceed. Both modeling activities normally proceed in parallel, although they will be presented sequentially here. In other words, steps 6 to 10 proceed in parallel with steps 12 to 16, even though the numbering scheme may not make that obvious.

[*Note*: The outline of the steps of building an information systems architecture in the diagram shows three dimensions that are modeled (the primary business-based dimension and two others). Although there always is the primary business-based dimension, there may be zero or more other dimensions in the model.]

THE ISA PHASE CHECKLIST (cont.)

6. *Creation of an ERD for each dimension.* The primary business-based ERD is mandatory (and may be the only ERD if there is only one dimension being modeled). There are three steps to be followed in the creation of the ERD: the identification of an entity; the definition of the entity; an identification of the relationship between entities. These three steps are normally iterated until a "clean" set of definitions results. A clean state is reached when: Nothing that belongs in a dimension is covered by more than one entity. In other words, everything belonging to the dimension is accounted for by one and only one entity.

   In addition, the definition of an entity must meet the qualifications for quality of definition.

   a. Does each entity definition meet the needs of the user? Of data processing personnel? Of management? If not, what must be changed?
   b. Are all the relationships identified? Have they been classified as to whether they are $1{:}n$, $m{:}n$, $1{:}1$, nonexistent, or otherwise? If not, what must be completed?
   c. For each user that participates in the dimension, does the ERD meet the users' needs? If not, what is missing? Are any users that participate in the dimension unable to contribute their needs? If so, why?
   d. Have relationships with other dimensions been specified? If not, why not?

   After steps (a) to (d) have been reiterated to the point where all data has been properly defined, the ERD is ready to be constructed. The ERD is simply the construction of the entities and relationships defined previously. If there are attributes that belong to an entity, they may be described at this point (or not; it is optional). *All* entities are defined. Care must be taken that all entities are at the same (or approximately the same) level of abstraction. If one or more entities are at a much higher (or lower) level of abstraction, the ERD should be reconsidered.

   OUTPUT: The output is an ERD for the dimension. Note that this step will occur for each dimension being modeled. Note that a corresponding activity for process definition is occurring at the same time.

7. *The DIS.* The data item set is created from the entities defined in the previous two steps. The DIS is nothing more than an expanded version of each entity. Data attributes and keys are added and the DIS is analyzed according to its "type" (i.e., by the different types of the entity that are defined).

THE ISA PHASE CHECKLIST (cont.)

a. What is the key for each DIS? Does the key meet *all* the needs of the dimension being modeled? What, in fact, are all the dimensions that must be satisfied for each key?

b. What attributes belong to each DIS? Are the attributes normalized? If not, why not?

c. What relationships are there to other dimensions? How are they defined at the DIS level?

d. Does an attribute appear multiple times in the same dimension?

e. What alternate keys are there? How are they specified?

OUTPUT: The output is a DIS for each entity belonging to the ERD. The DIS contains attributes, keys, nonunique keys, and so forth.

8. *Entity dimension maps.* After the DISs are created, the entity dimension maps are built (normally, the entity dimension maps are an informal part of the process, although they can be formally documented). This step must be done after *all* dimensions are modeled. In the case where all dimensions are *not* completely modeled, *at least* the DIS that will go into the entity dimension map must be modeled. Without *some* degree of completion across the model the appropriate data cannot be contributed to the entity dimension map.

   Not all DISs will go into an entity dimension map. Only those DISs that model the same entity should be combined. The output of the entity dimension map is the consolidated DIS. The consolidated DIS that results from the entity dimension map is the accumulation of data that results from combining the different, related DISs.

a. What dimension is represented by each dimension in the entity dimension map?

b. Should other dimensions participate in the map that are not available? If so, how are their requirements to be gathered?

c. Is more than one DIS from the same dimension participating in the map? If so, why?

d. What differences are there between participating DISs? How are the differences to be resolved?

e. What is to be done if one or more dimensions have not been fully defined? How will the requirements for that dimension be accounted for?

THE ISA PHASE CHECKLIST (cont.)

OUTPUT: An entity dimension map is created for all entities that are common to multiple dimensions. The map lists the entity and the different dimensions in which the entity participates. The output of the entity dimension map is the consolidated DIS. The consolidated DIS is nothing more than an amalgamation of the various data that belong to the entity that is collected from the different dimensions.

9. *Data views.* When the DIS and consolidated DIS are created, they must be verified in a bottom-up fashion. Data views (DVs) are created for different perspectives of each dimension. DVs are created strategically so as to include all appropriate views and to exclude redundant views, as much as possible.

   Once the DVs are created, they are matched against the DIS to see if all data that are in the DIS are there. It is normal to add some amount of detail to the DIS at this point. It is also normal to discover some small changes in the basic data model. It is not normal to discover *major* discrepancies at this point.

   If major discrepancies are discovered, some point of the design process done previously has not been correctly executed. Perhaps the scope of integration has not been properly defined or perhaps all the dimensions have not been outlined. A common mistake is for user views from another dimension to be merged with another DIS, creating wide discrepancies. The result of this process is a bottom-up verification of the top-down design that has been done so far.

   a. What data views have been strategically created to verify the DIS?
   b. What data views have been excluded?
   c. What amount of redundancy is there in the data views that have been created?
   d. What discrepancies have arisen in the comparison of the data view to the DIS?
   e. Have all data been specified in the DIS? Have the formats of the data been specified?
   f. Have commonly occurring data been separated from data of peculiar types? How? If not, why not?

   OUTPUT: The output of this step is a bottom-up refinement of the DIS created from the top-down design.

10. *Physical data model.* From the verified DIS that have been created, the physical model can be created. At this point, the major considerations of whether the

THE ISA PHASE CHECKLIST (cont.)

system is to be batch or on-line, operational or decision support, and so on, must be accounted for.

Of particular interest in physical design are the considerations of on-line performance and availability. Whatever the final physical disposition of the data, redundancy of the definition of data is minimal. There is a single data definition regardless of how many physical forms of data there are.

The physical model represents a culmination of all the different types and views of data that there are. It satisfies *all* the various needs.

a. Has the performance profile of the data been specified?
b. Has the availability profile of the data been specified?
c. What physical design features will be adopted to accomplish the performance and availability profile?
d. Have different modes of operation been identified? If so, what data will pass from one mode to the next? At what frequency?
e. Have all dimensions participated in the building of the physical model? If not, how will *total* physical requirements be satisfied?
f. What data base management system (DBMS) will be used? Does the data design fit well with the strengths of the DBMS? If not, why not?

OUTPUT: The output is a physical form of the data that can be defined to a data base management system.

11. *Basic process model.* (As noted previously, this step is done simultaneously with step 6.) The primary business-based BPM is defined, as well as the BPMs that describe other dimensions.

Like the ERD, this activity is normally iterated many times. In fact, this activity can be done in conjunction with the development of the ERD, since the level of modeling is the same. The first step in this process is the identification of a basic process as derived from the decomposition. Once the basic processes are identified, they are defined, using the same criteria for quality of definition as that used for entity definition. Next the cycle is identified. This activity is iterated until the cycle is defined with no ambiguities. All activities within the dimension should belong to one and only one basic process, as in the case of data and entities within the same dimension.

a. Have all functions in the dimension been accounted for by a basic process? Has any function in the dimension been accounted for by more than one basic process? Does the flow in the basic cycle account for all flow in the dimension?

THE ISA PHASE CHECKLIST (cont.)

b. Are multiple modes of operation represented in the ERD? If so, how do the cycles of each relate?

The creation of the BPM amounts to nothing more than a formalization of the output of the preceding step. The processing cycle is defined and is used as a model for further definition. Care must be taken that all processes are at the same level of abstraction. If there are major differences in the levels of abstraction, the resulting lower-level process models will reflect the problem.

OUTPUT: The output of this phase is a definition of the basic cycle of business that is being modeled.

12. *The FM and the DPM.* Each basic process is broken into a functional model. The FM represents nothing more than a more detailed outline of what activities occur and the cycle in which they occur. There is a distinction made between activities that are common and those that are unique, in much the same way that "type of" data is separated at the DIS level.

    From the FM, the DPM is created. The DPM represents an outline of *all* the functions that must be accomplished. The prerequisite and transformation data is identified at this point. One DPM is built for each FM.

    a. Have the common functions been separated from the unique functions? If so, how? If not, why not?
    b. For each process, what FMs have been created? Why?
    c. Have the prerequisite and transformation data for each FM been defined? If not, why not?
    d. What level of detail has been added to the DPM? Can program specifications be written from a DPM? If not, why not?
    e. Is there any function in the dimension that is not represented by a DPM? By an FM? If so, why?

    OUTPUT: The output is a detailed outline of the processing to be done. The detail includes algorithms, flow, relevant data that must be present for the process to be executed, relevant data that can be changed by the process, and so forth. "Pseudocode" can be written from the output of this step.

13. *Consolidation of the DPMs.* After the DPMs are built, those that are common are consolidated. This implies that all DPMs must be built for the consolidation to be complete. However, the DPMs that are recognizably common can be built for a dimension without building all the other DPMs.

THE ISA PHASE CHECKLIST (cont.)

The consolidation process produces a conglomeration of *all* the activities that are common. Usually, there is much overlap, but in cases where one process is common to another and there is a significant difference in processing, there needs to be reconciliation at a higher point in the process model, ususally at the decomposition or scope of integration level.

OUTPUT: The output of this step is a consolidation of all the processing to be done that is common. Both common requirements and unique requirements are identified here. The result is a process definition that satisfies all requirements whenever used but still allows for processing differences.

14. *The CDPM.* The result of the process model consolidation is the consolidated detail process model (CDPM). The CDPMs are verified together with the DPMs that are unique by means of process views. Process views are strategically created to ensure that the detail is correct. It is normal to have some amount of detail added to the DPM or CDPM as a result of PV verification.

    a. What DPMs go into the CDPM? What DPMs are excluded?
    b. Does more than one DPM from the same consolidated process participate in the same CDPM? If so, why?
    c. Do any DPMs diametrically differ from other DPMs in the same CDPM? If so, why? If so, how is the difference to be resolved?
    d. What PVs are selected for verification? Which are excluded? Why?
    e. Does any PV differ diametrically from the DPM or CDPM? If so, how is the difference to be resolved?
    f. Are common DPMs easily related to all the PVs that belong to it? If not, why not?

    OUTPUT: The output of this step is a function refinement of the processes that meets the detailed needs of users at the lowest level.

15. *Program specification model.* Once the CDPMs and DPMs are verified by the PVs, the program specification model is created. The program specification model is the documentation from which code can be written. Prerequisite data and transformation data are specified as well as detailed processing instructions. Where code is common it needs to be specified as such.

    OUTPUT: The output of this step is documentation from which program specifications can be written.

THE ISA PHASE CHECKLIST (cont.)

16. *Final verification.* Once the physical data model and the program specification model are created, they are "married." The marriage represents a verification that all data that are needed are present and that all processing that is needed is specified. After the marriage is complete, physical data base design can begin and programs can be written.

    A word needs to be said about changes to the model that has been created. The user's environment is a changing world. It is normal for some amount of unanticipated change to occur. In theory the change should be filtered back through the modeling process from the first step downward. In theory the change that is wrought at each step should be small if the model has been created properly.

    The problems occur when entirely new dimensions are introduced; the scope is changed to include (or interface to) entirely different modes of operation; or the business undergoes radical change. Of course, anticipation of these events *may* have been built into the model, but it is entirely possible that these events could not have been anticipated.

    When an unanticipated, catastrophic change occurs, it is not unreasonable that the impact is also catastrophic. But it is inexcusable to allow a small change to produce a catastrophic result. A well-built information systems architecture is able gracefully to withstand small changes with the minimum amount of impact.

2

Entity-Relationship Diagrams: Top-Down Design

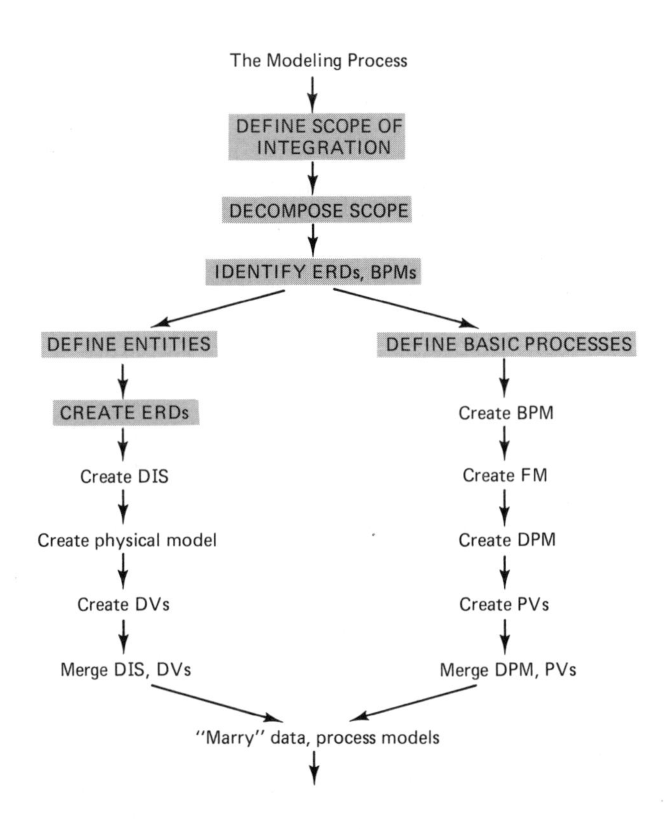

If the scope of integration has not been firmly established, the remaining design steps should not proceed. When the scope has not been established, doing further design work is risky because entire sections of design are subject to large amounts of change as the scope expands and contracts. The entire design is based on the framework outlined by the scope of integration. A proper definition of the scope of integration should be:

- Brief
- Simple
- Definitive

For example, if the scope of integration were "all systems of the Bank of Tomorrow," the description would be brief and simple, but not definitive. Does the scope include administrative and operational systems? decision-support systems? Archival systems?

Perhaps a better scope might be "the operational systems of the Bank of Tomorrow." Even at that, does the definition include commercial and retail systems? Wholesale systems? Data processing service bureau systems? Perhaps the scope should be refined even more, to "the financial banking retail systems of the Bank of Tomorrow." At this point the scope is probably meaningful. However, the importance of the scope of integration and its impact on system modeling is obvious when the three scopes mentioned here are considered.

1. All systems of the Bank of Tomorrow: Payroll, Personnel, Rental Property, Loans, Trust, Financial Decision, Tax Records, Service Bureau, Correspondent Bank, Savings, DDA, Audit Control, Insurance, Retirement Benefits, General Ledger
2. Operational systems of Bank of Tomorrow: Financial Decision, Loans, Tax Records, Savings, DDA, Trust Correspondent Bank, General Ledger
3. Financial banking retail systems of Bank of Tomorrow: Loan, Savings, DDA, Trust

The model of data and processes for scope 1 will be a far cry from the model for data and processes of scope 3. If the scope has not been clearly stated and well considered, the resulting design effort is jeopardized and large amounts of time will probably be spent in struggling with what is and what is not within the scope of effort both at the moment of top-down design and later, when questions arise as to why something is or is not in the model. Because of the importance of the establishment of the scope, it can clearly be stated that *no design efforts should proceed until the scope of integration is established and management understands and formally acknowledges the cost of changing the scope at a later time.*

Establishing the scope of integration may be made easier by building a map of the global scope of integration. Different modes of operation, such as decision-support and operational systems, belong in different scopes. Other major divisions of systems, such as administrative and operational, may be included on the map.

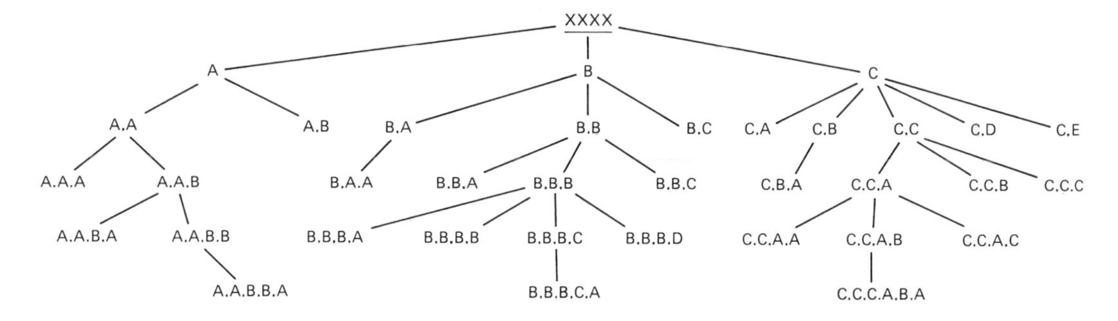

Figure 2.1 Scope of integration is identified by *xxxx*. The diagram reads: *xxxx* is made up of A, B, and C; A is made up of A.A, A.B, and A.C; A.A. is made up of A.A.A and A.A.B, and so forth.

ONCE THE SCOPE IS ESTABLISHED

Once the scope of integration is established, it usually is characterized by a system name—something that brings to mind the scope to user, manager, and system architect. Then the scope is decomposed into its organizational components, usually along lines of the organization chart that supports the decomposition. Decomposition starts at the top and proceeds downward, as shown symbolically in Fig. 2.1.

There are many ways that the decomposition process shown in Fig. 2.1 can be done, some of which are useful and some of which are not. To better illustrate a useful decomposition of a scope of integration, consider the following two scopes: the financial systems of a bank, and the systems of a manufacturer. The decomposition of the scope of the financial systems of a bank is shown in Fig. 2.2. The decomposition of the scope of integration for a manufacturer might look like Fig. 2.3.

Figure 2.2 Scope of integration: the financial systems of a bank.

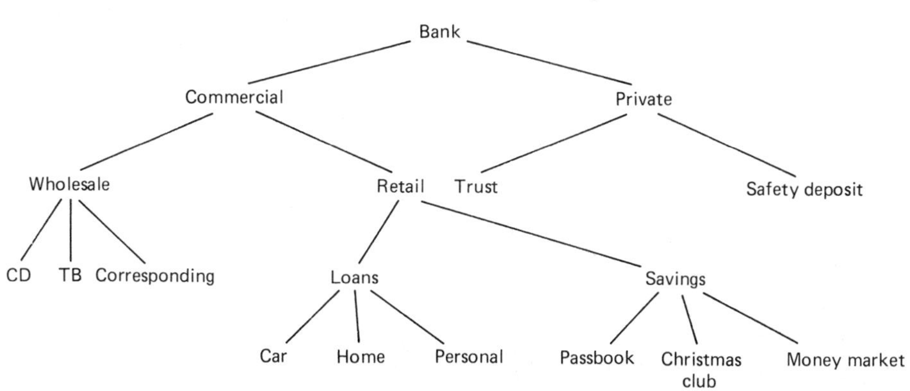

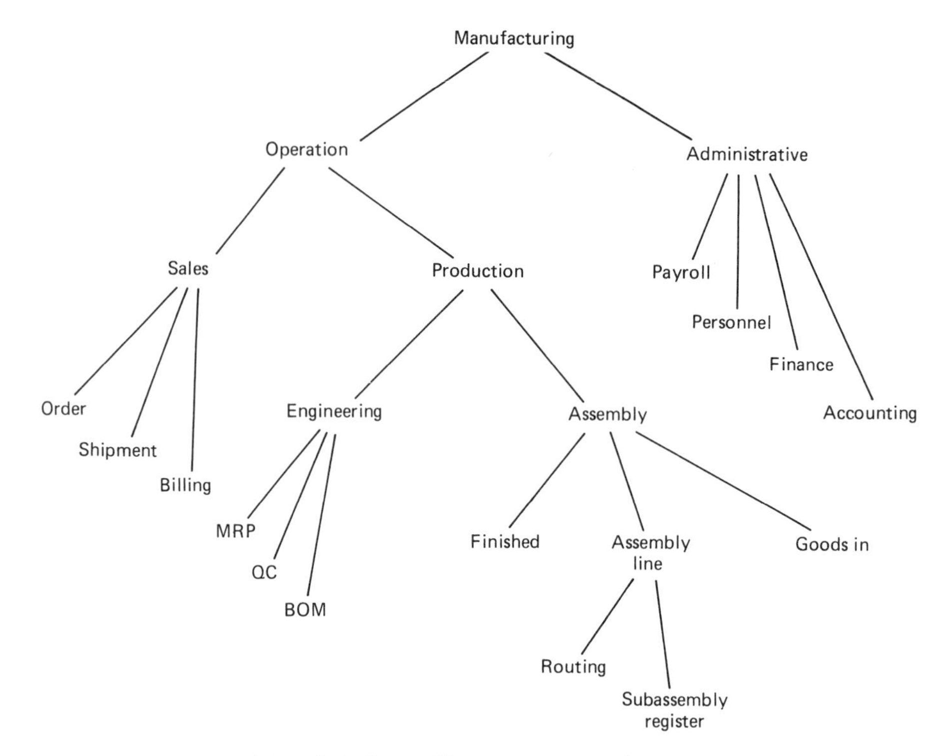

Figure 2.3 Scope of integration: a manufacturer.

There are a few points of interest in the scopes of integration depicted in Figs. 2.2 and 2.3. The first point of interest is at what level the decomposition process stopped. In both cases the decomposition stops at the point of lowest functional distinction (sometimes called the "primitive" or "atomic" level). The designer knows that the point of lowest functional distinction is reached when the process cannot be subdivided into a lower set of distinctive functions (i.e., the function loses its distinction if further divided). For example, the process of accepting money is nondistinctive because it could be for loans, savings, or checking. It could also be for making change or for replenishing another teller. There are many reasons why a bank teller accepts money. So the process of decomposition should not go any further than the point at which distinctive activities are separated.

Decomposition is best understood in terms of examples. Consider the decomposition of commercial banking. It can be subdivided into a lower set of functions (wholesale, retail) and each of these activities has functions that are unique to itself, so commercial banking is subdivided.

Consider passbook savings. Should it be subdivided into a lower set of activities? The first questions is: Can it be subdivided at all? The answer is yes, it can be broken up into teller activity, interest calculation, customer activity, and so forth. But these lower sets of activities can be found throughout the bank. They do nothing to set passbook savings management apart distinctively from other activities, so it is not appropriate to decompose passbook savings further. The function can be further subdivided but not without losing its distinctive qualities.

The second point that is noted concerning the decompositions shown in Figs. 2.2 and 2.3 is that the decomposition says nothing about the business of the organization, but describes how the *organization* is structured. The decompositions shown describe the organization of the bank or manufacturer. They say *nothing* about the customer, for example. This distinction between the business of an organization and the organization itself is important in that a system model must be based on the *business* of an enterprise, not an organization chart of the enterprise, even though the first steps of decomposition usually begin with a decomposition of the organization chart.

The third point of interest concerning the decompositions is that in one case of decomposition, operational systems are mixed with administrative systems and in the other case they are not. For all practical purposes (because operational systems represent one mode of operation and administrative systems another mode, they should be separated, as shown in Fig. 2.4. The model that results when operational and administrative systems are mixed tends to be very confusing and misleading. A better ap-

Figure 2.4 Scope of integration: a manufacturer. Operational and administrative systems are separated and a new set of scopes are defined.

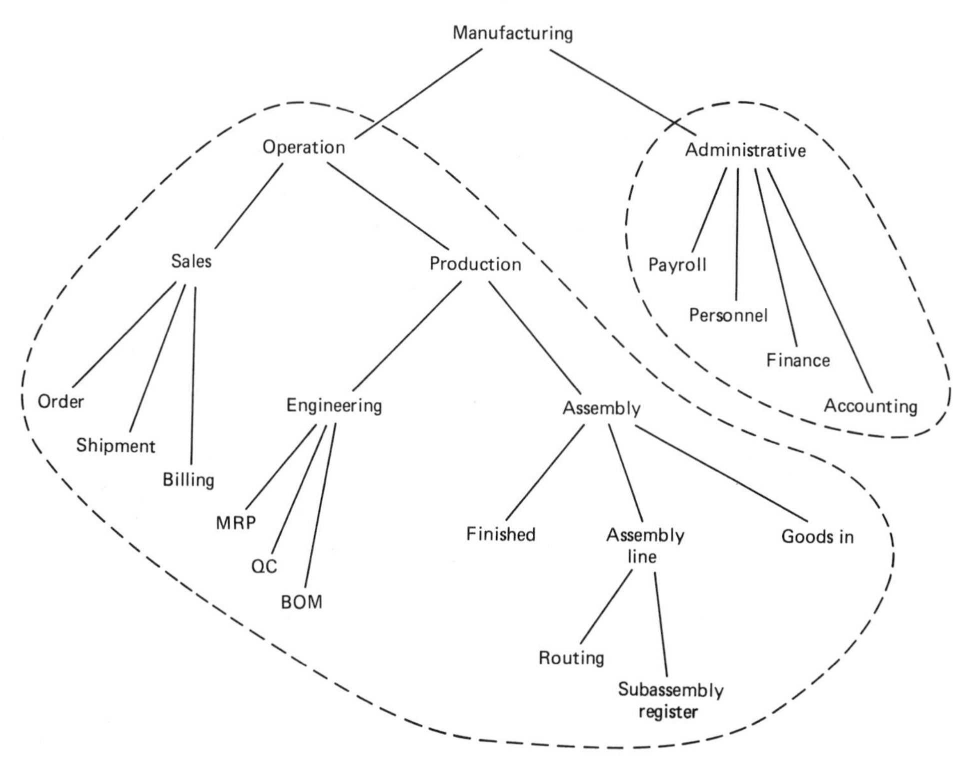

proach is to construct two models, one for operational functions and one for administrative functions. This requires, of course, that the scope of integration be revised if the original scope included both modes of operation. In general, different modes of operation—administrative, operational, decision support, and so on—should not be mixed within the same scope of integration. If modes of operation are mixed, inevitably there is confusion.

Once the scope of integration is established, it may be useful to identify all dimensions within the scope that will be modeled. For example, if the scope of integration is the financial systems of a bank, several dimensions will need to be modeled —the primary business-based dimension, the auditing dimension, the organizational structure dimension, and so on. To show how these dimensions relate, a global ERD map may be created.

ENTITY SELECTION

After the scope of integration has been established and has been decomposed to the proper level and the various modes of operation within the scope have been identified, the next step is to examine the business that is being done at each level within the mode of operation. For example, as shown in Fig. 2.3, the system architect in the manufacturing environment analyses what is happening at the engineering quality control level and determines that a part number is being manufactured and inspected. There are certain criteria that determine whether or not the part meets specifications. Then the systems architect analyses inventory control and determines that parts are shipped, received, counted, and so on. After a similar analysis for each of the functions at the lowest level of functional decomposition, the most common entities and processes are selected to become part of the ERD and the BPM, as shown in Fig. 2.5. In the case of manufacturing, the part becomes an entity and the manufacture of the part becomes a process. The first ERD that is created is the one that represents the primary business of the system. It is called the *primary business-based ERD.*

Once the entities are selected for an ERD, the next step is to define the relationships between entities. The relationships can be 1:1, 1:n, m:n, or nonexistent. If a relationship is 1:n, many entities may exist when one other related entity exists. Where the relationship is m:n, any number of entities may exist in relation to any number of other entities. To illustrate these relationships, consider the entities from a bank —customer and account—and the entities from a manufacturer—part and supplier. Figure 2.6 illustrates these relationships. A customer is seen to be able to have multiple accounts, but any account can be owned by only one customer. In the part/supplier relationship, a part can have multiple suppliers, while a given supplier is able to provide many different parts. These relationships are depicted by a single- or a double-headed arrow. A verbal description of the relationship may be provided for further documentation.

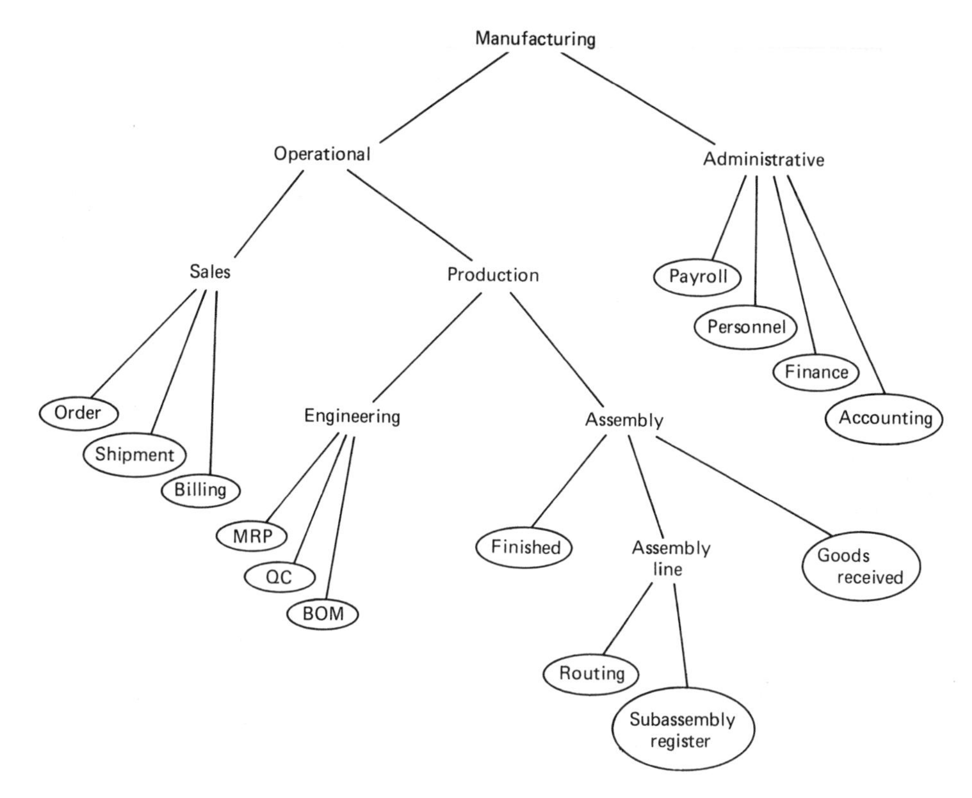

Figure 2.5 Scope of integration: a manufacturer. At the lowest level of decomposition, common data and processes are identified. This synthesization is the basis for identifying the primary business based data and process model.

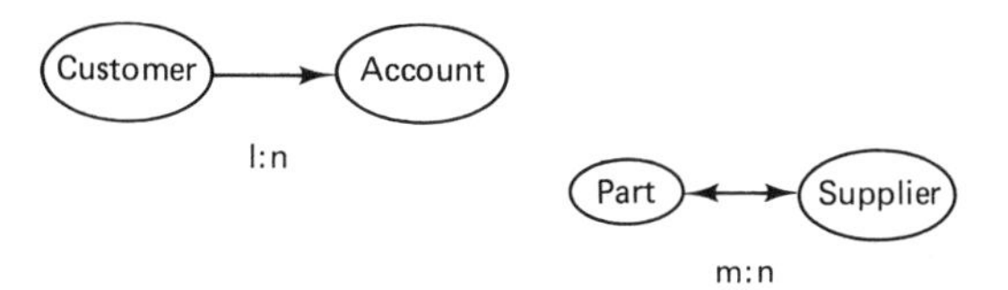

Figure 2.6

As an example of the entities and their relationships that are extracted from the banking and manufacturing decomposition, see Figs. 2.7 and 2.8. The banking ERD states that each account is domiciled at a unit (a branch, the main office, etc.). There can be multiple accounts per unit but only one unit per account. An account may have an associate (husband/wife, father/son, etc.). A customer may have multiple accounts, but an account serves only a single customer. There is at least one service agreement for an account (checking, bankcard, loan, etc.) and a service agreement may exist for more than one account. Activity exists according to a service agreement

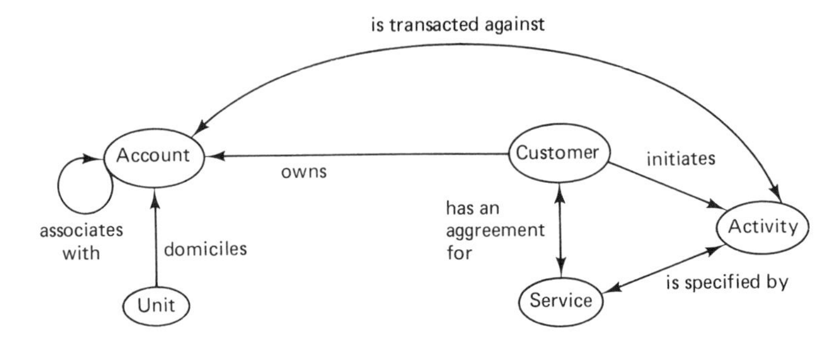

Figure 2.7 Typical banking entity-relationship diagram (ERD).

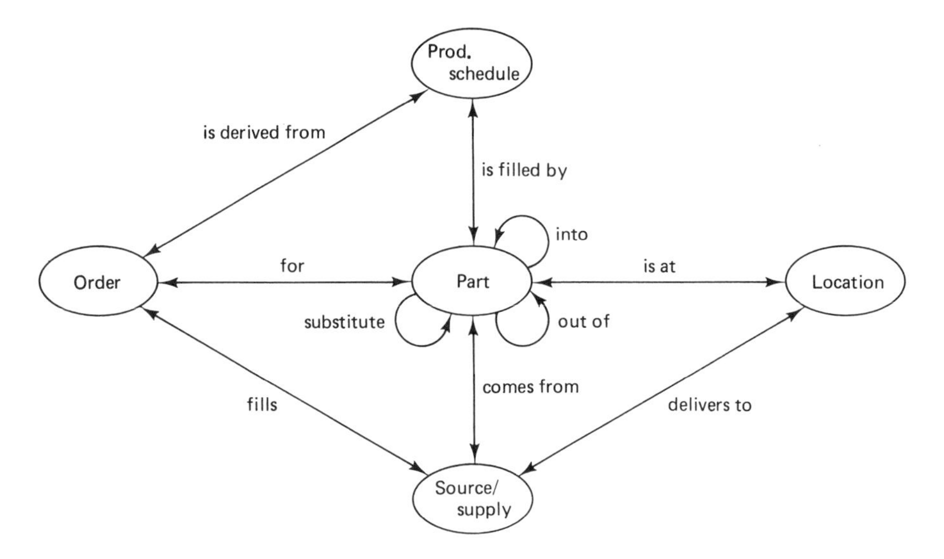

Figure 2.8 Typical entity-relationship diagram (ERD) for the manufacturing environment.

and a service agreement may be executed by multiple activities. A customer initiates one or more activities, but an activity is initiated by only one customer. Finally, activity is recorded by account, and activity can be transacted against multiple accounts.

The manufacturing ERD shows that there are recursive relationships from part to substitute part, into (bill of materials implosion), and from (bill of materials explosion). A part can exist in multiple locations and a location can have multiple parts. A part can be supplied by more than one source, and a source can supply multiple parts. Orders can exist for multiple parts and a part can be on multiple orders. The production schedule is filled by more than one part and a part can exist on more than one production schedule. The production schedule is derived from multiple orders and an order is filled by multiple production schedules. An order can be filled from multiple sources and a supplier can fill multiple orders. Finally, a supplier can deliver to multiple locations and a location can be served by more than one source.

The bank or manufacturing ERD examples are useful in describing a basic difference in the two environments. Even though the banking environment has many vagaries, the essential business of the bank is relatively straightforward, and that is demonstrated by the relative simplicity of the bank's ERD. In contrast, the manufacturing ERD is rich with relationships, which reflects a dynamic environment.

WHAT THE ERD DOES NOT CONTAIN

While the ERDs as presented are useful, even more instructive is what the ERDs do not contain. The purpose of the ERD is to focus on the business of the corporation, at the highest level of abstraction. Anything not meeting that purpose does not belong in the ERD. Certainly, there are other valid views of a corporation, and those views need to be modeled at the ERD level, but the first and most important data model that is built is the ERD that addresses only the primary business of the corporation.

For example, not found in the primary business-based banking ERD is the banking organization structure. Such a structure might have been included (but was not), as shown in Fig. 2.9. After all, when transacting daily business a customer sees the bank as a cohesive entity, not in terms of the organization chart. The organization chart represents the organization, not the business of the organization, and thus does not belong in the primary business-based ERD.

As an example of another type of entity that does not belong in the primary business-based ERD, consider Fig. 2.10. In this case it is shown that demand stems from a customer. Although this is true, as it is true that an ERD can be created at another level (a higher level of abstraction), the primary business of the bank is at another level of conceptualization (a lower one) than such entities as supply and demand. Demand is at a higher level of abstraction than the everyday business of the bank and does not belong in the primary business-based ERD.

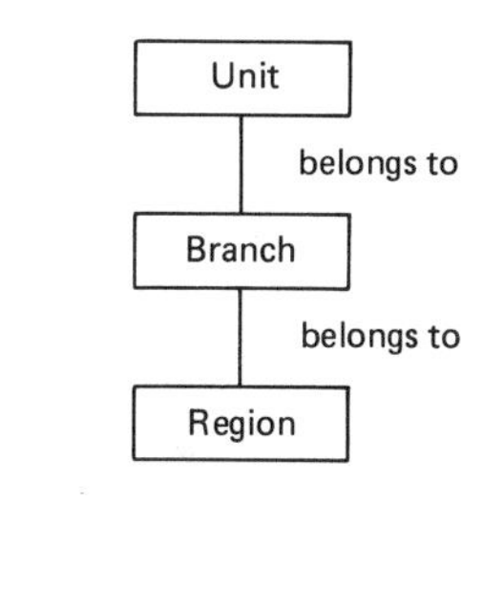

Figure 2.9 Organization breakdown of the bank—*not found* in the primary ERD of the business of the bank.

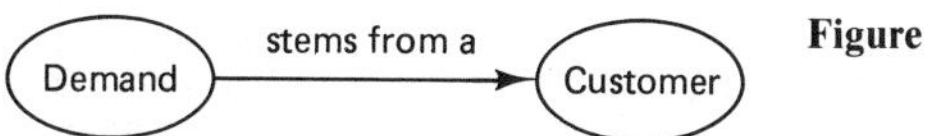

Figure 2.10

Another type of relationship not found is illustrated in Fig. 2.11. A customer is shown activating savings payments, loan payments, savings withdrawals, and loan extensions. Although each of these entities and relationships is valid, they do not represent the highest level of abstraction. The associated chain of abstraction shows that savings and loan activity are all forms of the same things—banking activity. Thus the primary business-based ERD shows the customer relating to activity, not all the various forms of activity.

Another type of relationship that does not appear in the primary business-based ERD is shown in Fig. 2.12, where the activity of *initiating* a relationship (signing up for a checking account, closing a Master Card, etc.) is seen to be only an adjunct of the primary business of the bank. Certainly, initiating service relationships is necessary, but it is not how the bank makes a profit. The bank makes a profit by fulfilling the service agreement that it makes with its customers (for a price, of course), not by creating service agreements. Thus the initiation of services does not belong in the primary business-based ERD.

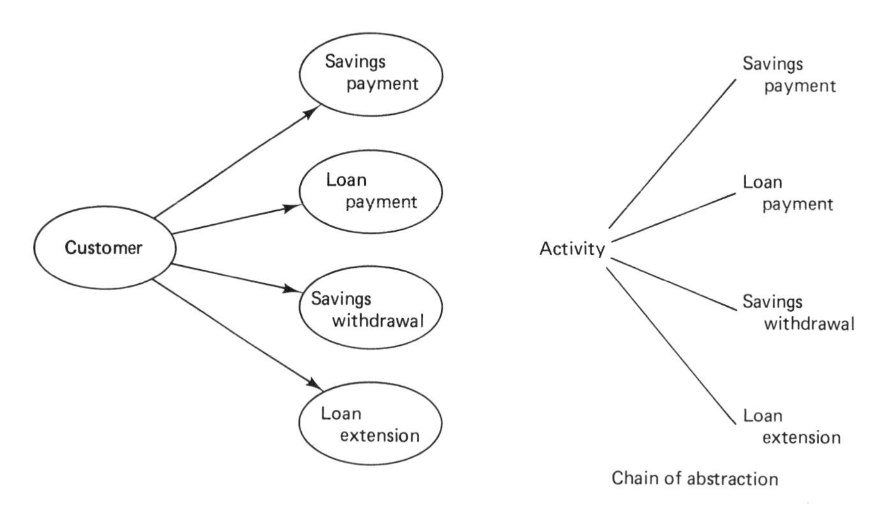

Figure 2.11

Figure 2.12

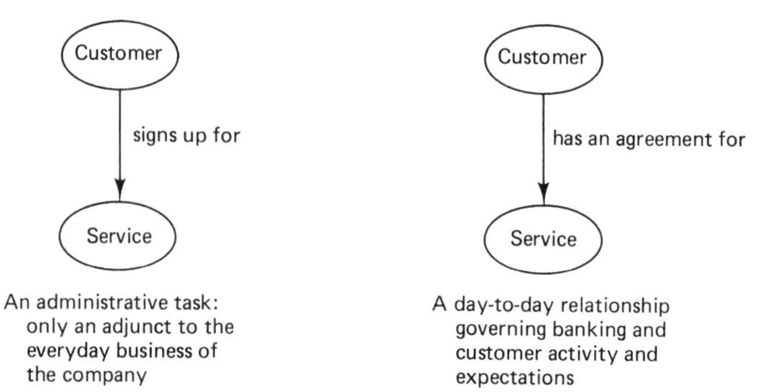

The manufacturing ERD has several forms that are of interest. Consider how recursion is represented. It is *not* represented, as shown in Fig. 2.13. Even though a part number goes into an assembly and a part number has substitute parts, the representation is at the highest level of abstraction where there is only one part. An assembly is just another form of a part, as is a substitute part. This abstraction is shown in Fig. 2.14.

Still another form of interest is the combination of source/supply found in the manufacturing ERD. The source/supply can be taken to mean a supplier (i.e., an external supplier—Jones Hardware) or an assembly line (where a part is made from raw goods, other parts, etc.). There certainly are major differences between an external supplier and an assembly line (and at a point closer to the implementation of the system those differences must be dealt with), but at the highest level of abstraction, from the view of the manufacturing process, the two types of sources are equivalent. For example, if assembly line XYZ needs a part, the line could not care less whether the part comes from an outside supplier or another assembly line—all line XYZ knows is that it needs the part.

The primary business-based ERD then has entities and relationships that satisfy three criteria:

- Are the entities at the highest level of abstraction that is appropriate to the scope of integration?
- Do the entities and relationships reflect the basic business of the company?
- Are the entities at a level beyond the scope of integration?

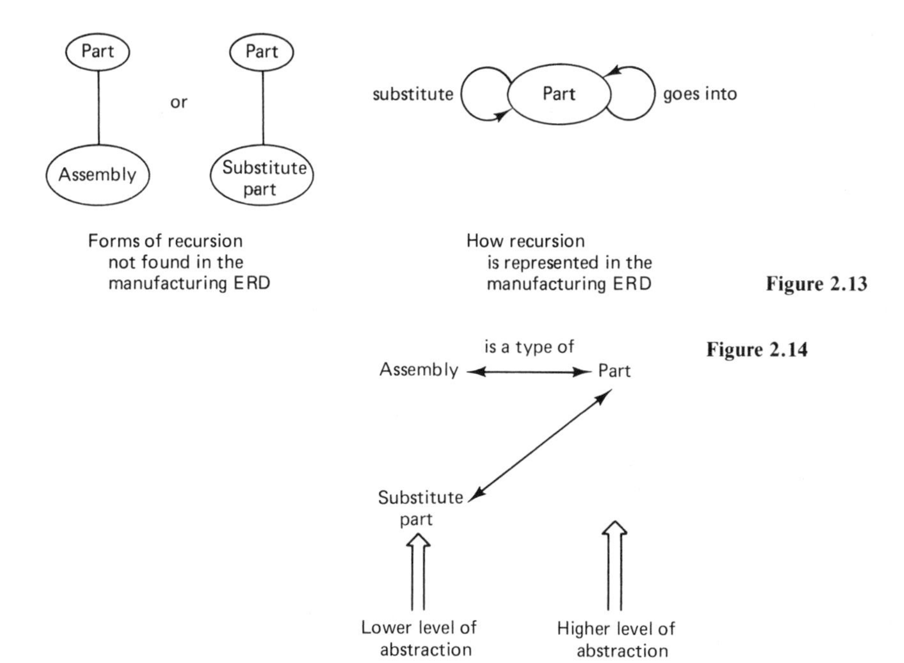

Figure 2.13

Figure 2.14

After the primary business-based ERD is built based on the scope of integration, the organization can *separately* build other ERDs, which reflect such things as the organization structure, the administrative activity, the tangential business, and so on.

FILLING THE SCOPE OF INTEGRATION

The first step in building systems suitable for information engineering is the definition of the scope of integration. The next step is to define the primary business-based ERD, which encompasses much of the scope of integration. After the primary business-based ERD is defined, other appropriate ERDs are defined until the entire scope is defined. This progression is depicted in Fig 2.15.

To be complete, the scope of integration must be defined completely, but in practice it may be *years* before all ERDs are built. In practice, the primary business-based ERD and two or three other important ERDs are usually constructed, leaving some portions of the scope of integration undefined. As long as the undefined portions are truly nonessential, or are deferrable, this somewhat less than ideal state is at least temporarily livable. In theory, the entire ERD should be defined; in practice the scope is usually so large that some less significant parts remain undefined.

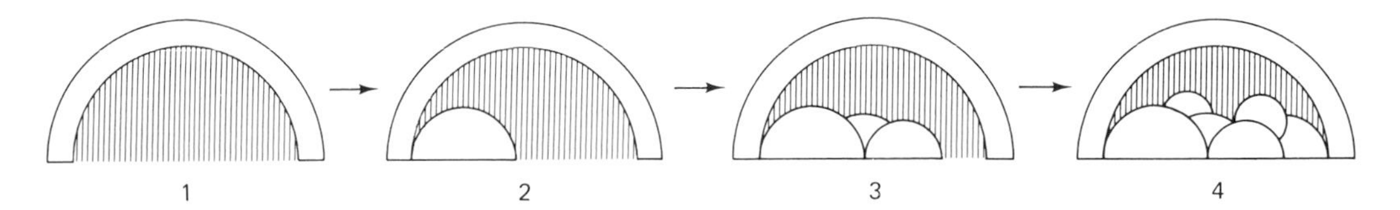

1. The scope of integration is defined.
2. The primary business-based ERD is defined.
3. Other ERDs are defined.
4. As time passes the scope of integration is eventually fully defined by ERDs.

Figure 2.15

ENTITY DEFINITIONS

Once the primary business-based ERD is defined (i.e., entities identified and relationships labeled), the next step is to define each of the entities. As a rule, the entities ought to satisfy (at the least) the following criteria:

1. Does the entity definition encompass *all* that it ought to encompass?
2. Does the entity definition encompass *only* what it ought to encompass?
3. Is there any occurrence of an entity that is not covered by a definition?
4. Has every term used in a definition been defined?

5. Is there any occurrence of any entity that is covered by more than one definition?
6. Are any two entities a form of a higher entity definition?
7. Are the definitions readable and understandable (especially to the user and management)?
8. Does the user concur with the definition?

As an example of the importance of each of these criteria for satisfactory definition, consider the following definitions of a banking customer: "A customer is a person who uses services." Applying the first criterion—Does the entity encompass *all* that it ought to encompass?—brings to light the fact the companies are also bank customers, but a company is not a person. The definition is refined to—"A customer is a user of services"—which encompasses both people and companies.

Now the question is asked: Does the definition encompass *only* what it ought to encompass? What about a pedestrian who wants to use the restroom? Under the definition as given the pedestrian is encompassed, but probably should not be. The definition is now refined to—"A customer is a user of financial services"—which disqualifies the pedestrian.

The next criterion is now applied: Is there any occurrence of an entity that is not covered by a definition? The question brings to light the fact that a person holding a credit card that has yet to use the card is a customer of the bank, but not according to the definition. The definition is now revised: "A customer is a user or potential user of financial services according to a prior service agreement."

The next criterion becomes relevant: Has every term used in a definition been defined? In this case "prior service agreement" has not been defined. A definition is drawn up for a service agreement to read: "A service agreement is a covenant between the bank and an external party governing the transaction of financial acitivity."

Now the definition process progresses and it is noted that there often is an association between two customers, such as husband and wife, or partners. So a new entity is defined: "An associate customer is one that has a financial relationship with another customer."

Now the next criterion is applied: Is there any occurrence of any entity that is covered by more than one definition? It is noted that a customer may at once be a customer and an associate customer. In addition, it is noted that an associate customer is just another form of a customer, thus filling the next criterion. So a recursive relationship to customer is formed, and the entity "associate customer" disappears.

The definition of a customer is passed to the legal department and the definition now reads: "A customer is a party to fiduciary and financial fluctuations in accordance with a prior mutually binding contract in compliance with statutory regulations." The system architect examines the definition and decides that although it may satisfy legal requirements, it does not satisfy everyday common usage requirements for communication. Thus the earlier definition—"A customer is a user or potential user of financial services according to a prior service agreement''—is kept for every-day use.

Finally, the user and management are asked to review the definition to ensure that it meets their needs. Throughout the definition process, the user must have played an active role. Without user participation the process of defining entities becomes a backroom ''techie'' exercise and risks bearing no resemblance to reality.

ITERATING THE ERD PROCESS

It is normal to iterate the process of identifying and defining entities and relationships. This iterative loop is shown by Fig. 2.16, where it is seen that there are three major steps: identifying the entity, identifying the relationship, and defining the entity. It is *normal* during the rigorous definition process to find that there are loose ends: incomplete definitions, entities that need to be consolidated or included, and so on. Whenever a definition causes a change in the data model, the process of modeling must be reiterated until the final set of entities, relationships, and definitions are consistent and satisfy the definition criteria.

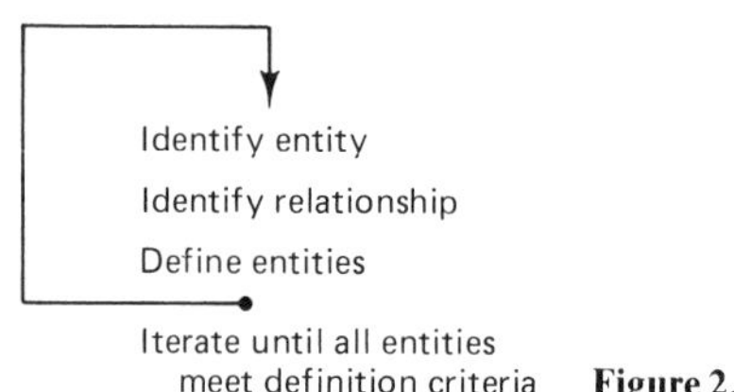

Figure 2.16

ENTITIES AND DEFINITIONS

The definition of an entity should be consistent throughout the mode of operation within the scope of integration. That is why the primary business-based ERD is the first and most important ERD to be constructed. As other ERDs are defined, they normally build on previously defined ERDs. But what happens if the scope of integration chosen is *very* wide and crosses broad functional ground (i.e., encompassing many modes of operation)?

For example, suppose that a bank makes money from three types of activities:

- Financial activity (normal banking activity)
- Service bureau activity (where the bank processes payrolls, insurance, benefits, etc.)
- As a landlord (the bank leases out buildings that it owns)

Suppose that the scope of integration was defined to include *all* these activities. Note that each activity represents an entirely different function. Based on the functional differences, the three functions can be said to represent different modes of operation. Should a definition fit *all* three different business functions? Consider the following definitions:

- *Customer (financial):* party with whom financial service agreements are made
- *Customer (service bureau):* user of data processing services
- *Customer (leasing):* party renting space from the bank

It makes *no sense* to have a definition consistent across broad, differing functional lines that goes across several modes of operation unless the system is being built for the aggregation of the businesses of the bank, such as for a system that determines the bank's balance sheet. Instead, the scope of integration should be divided into realistic boundaries.

But consider the definition *within* functional lines (i.e., within the same mode of operation). Should the bank have a single definition of a customer as far as loans, savings, trust, credit card, and DDAs are concerned? By all means. The definition of customer should be consistent across the same functional area—in this case, financial services.

CONSISTENCY OF DEFINITION

The importance of consistency of definitions across the same mode of operation is illustrated by an example. Consider something simple such as a loan. On the surface a banker or customer immediately assumes that he or she knows what a loan is. But consider how a loan can be classified:

- Car
- Home
- Signature
- Commercial
- Secured

At this point there is little dispute as to what a loan is. But consider the different statuses that a loan might go through.

- Default
- In progress (committed)
- Second party (joint holdings)
- Applied for but not approved
- Litigated

It is questionable in the foregoing cases when a loan is a loan (or, in fact, if a loan is a loan). The definition becomes even more confused when other types of loans are considered.

- In house, interdepartmental
- Subsidiary loans (by a holding company)
- Guaranteed, defaulted loans

Now the understanding of what is a loan is really complex, especially when viewed from different perspectives, such as the following:

- The stockholder
- The IRS
- The board of directors
- The federal auditors
- The managing committee
- The customer
- Public relations

As an example of information for which there is a need for a precise definition of a loan, consider the following questions:

- In terms of outstanding loans, where is a loan in default accounted for?
- In terms of an in-house loan, how is it viewed by the IRS?
- In terms of the stockholders, what if a subsidiary has a different definition of a loan and consolidated report is to be issued?
- In terms of the federal auditors (and a customer), should a loan applied for and committed, but not paid, be counted as a loan?

And so on. It is clear that a well-thought-out, consistent definition is mandatory. The data that will be running through the systems based on the model being created are *first* dependent on their definition for meaningful processing to occur. Without a proper definition, the implementation of a system is of little value.

IN SUMMARY

The first step in building an effective business-based systems model is to determine the scope of integration. Once the scope has been determined, it is decomposed, usually along organizational lines, until the lowest level of distinct functional processing is determined. The various modes of operation within the scope are outlined based on the decomposition. From this decomposition the primary data and processes that are common to the most functions are extracted, and the primary business-based model is created. The first step in the building of ERDs is to identify entities. Then relationships between the entities are created. Finally, definitions for the entities are created, under a rigorous set of criteria of data definitions. The output of this first step of system modeling is the ERD.

PROJECT STUDY

ABC has decided that the only long-term solution to the maintenance, development, budget, and user dissatisfaction problem is to integrate their systems. Although there are some short-term technical improvements that can be made, ABC looks at the problem from a larger, long-term, corporate viewpoint. ABC realizes that a certain amount of pain and a certain number of changes must be accepted with today's systems. The current mess is going to remain a mess for the life of the existing systems. But ABC does not want to repeat the mistakes of the past and looks strategically at tomorrow's data processing world with an eye toward controlling data processing, as it becomes increasingly more important to the bank.

ABC creates a new function within the organization—the information control officer (ICO). Although the ICO has a small staff, the ICO is given blanket veto power on any phase of development or maintenance of any current or future project.

The ICO gathers a team of people (whose makeup changes from time to time). This team, made up of users and applications personnel, is called the *information architecture* (IA) *team*. The IA team selects as its first task that of leading ABC out of the quagmire by defining the scope of the task—the scope of integration. There is much discussion as to what does and does not belong in the scope. One philosophy is to include all aspects of ABC's business. This would include financial systems, rental systems, software royalty systems, software services, processing services, and other miscellaneous activities. Together with this consideration is what to do with administrative systems: Do they belong with ABC's operational systems?

The initial reaction of the IA team is to define the scope so that all the business activities will fit within the scope, since after all, the businesses are part of ABC. But another philosophy is to limit the scope to the "bread and butter" systems of ABC—the financial systems that form the basis of ABC. This limitation of scope is thought to provide a more "doable," more realistic basis for the building of future systems.

Based on the rationale that the IA team must consider what has the greatest chance of succeeding, the IA committee decides to limit the scope to ABC's financial systems, but to include the bank's administrative systems as well. The scope is now appropriately formally defined by the IA team and is transmitted to management. The next step undertaken by the IA team is to functionally decompose the scope of integration. The functional decomposition is done along the lines of ABC's internal organization. Figure 2.17 illustrates the decomposition done by the committee.

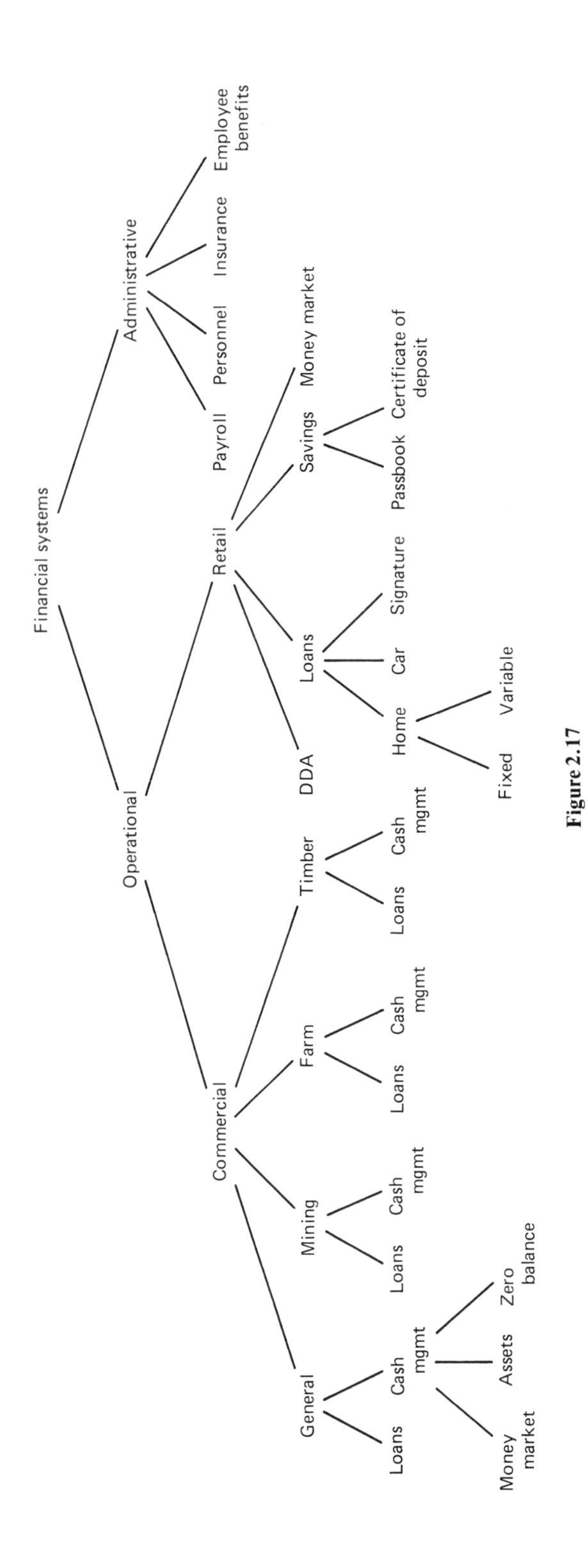
Financial systems
Operational
Administrative
Commercial
Retail
Payroll
Personnel
Insurance
Employee benefits
General
Mining
Farm
Timber
DDA
Loans
Savings
Money market
Loans
Cash mgmt
Loans
Cash mgmt
Loans
Cash mgmt
Loans
Cash mgmt
Home
Car
Signature
Passbook
Certificate of deposit
Money market
Assets
Zero balance
Fixed
Variable

Figure 2.17

After the decomposition is done it is decided to construct the several ERDs that represent the data inside the scope—one for administrative systems and one for operational systems. It is suggested that operational systems be divided into two separate ERDs: a commercial ERD and a retail ERD. But that suggestion is discounted because there is so much commonality of function among the two types of business. So an ERD is to be constructed that reflects *both* commercial and retail requirements.

The entities that have been identified are for the primary business-based ERDs: customer, account, activity, and service agreement. The processes that are common to the ERD are acceptance of risk, transaction (payment, deposit) against risk, and risk-related calculations. The definitions of the entities and the relationships among the entities are now addressed by the IA committee. As an example of the ERD that is being discussed, refer to Fig. 2.18. The entities are then defined, using the criteria for the quality of definition. As an example, the definition of activity is: "An activity is a recorded exchange of funds, equity, or assets in accordance with an account service agreement." All other definitions are established similarly.

1. How long should the activities of the IA team, as described, take? Six months? Six weeks? Two years? Does the length of time matter? Why?
2. What would have happened if ABC has chosen to model *all* their businesses under the same scope of integration?
3. As the entities were defined, is it possible that "activity" was the first entity to be defined? Why not?
4. Suggest changes to the ERD. Are all relationships properly defined?
5. Why isn't a bank branch a part of the ERD? Isn't it a part of ABC?
6. Aren't there many financial activities that occur that are not part of the processes selected?

Figure 2.18

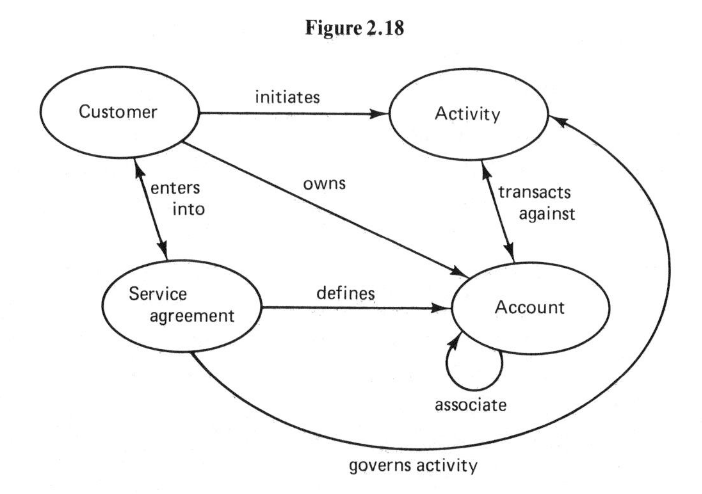

7. Describe how the entities and processes were selected from the decomposition.
8. Describe a way that decomposition could have been done on other than an organizational structure.
9. Why doesn't the decomposition consistently go to the same levels? Why isn't the structure that results from the decomposition symmetrical? Can it ever be symmetrical? Under what circumstances?

EXERCISES

1. (a) If the scope of integration is not complete, design work should not proceed. Why?
   (b) What happens if design work does proceed? What is at risk? What resources are being risked? At what level (i.e., what magnitude of resources) is the risk being taken?

2. (a) What should go into the definition of the scope of integration?
   (b) How long should it take to create a scope of integration?
   (c) How should the scope be documented? Where should it be documented?

3. (a) Who should determine the scope of integration? What is the result if any one party is left out?
   (b) Who should not participate in the definitions of the scope of integration? Why?
   (c) Who should be allowed to change the scope of integration? When? Under what circumstances?

4. (a) What is organizational decomposition?
   (b) How does it differ from process decomposition? From data decomposition?
   (c) Decompose the following businesses functionally:

   (1) A large, multinational bank
   (2) An independent insurance agent
   (3) A manufacturer of marbles
   (4) A distributor of vegetables and eggs
   (5) A sports franchise

5. (a) How should entities be selected from the organizational decomposition?
   (b) How can a primary business-based entity be distinguished from a nonprimary business-based entity?
   (c) How many entities belong in the primary business-based ERD?
   (d) How can the designer determine if all the entities that should be in the primary business-based ERD, in fact, are there?

6. (a) What is the importance of the $1{:}n$ or $m{:}n$ identification of a relationship?
   (b) Is there ever an entity that has no relationship to another entity and still belongs in the ERD?
   (c) Are there relationships other than $1{:}n$ and $m{:}n$? If so, what are they?

7. Why is the highest level of abstraction that is appropriate to the scope of integration proper for the primary business-based ERD?

8. There are plenty of valid views of a corporation other than the primary business-based views.

(a) Name five or six of them.

(b) Why are they *not* appropriate for inclusion into the primary business-based ERD?

(c) To what are they appropriate?

(d) Should they be discarded? Are they invalid?

9. (a) Describe the process of filling in the scope of integration.

(b) By when should it be filled in? Is it possible that some of the scope will never be filled in? Is this harmful? In every case?

10. Using the criteria for entity definitions, define the following entities:

(a) A part

(b) A bank account

(c) An insurance policy

(d) A savings deposit

(e) An order

(f) A substitute part

(g) A loan

11. The "width" of a definition can be said to describe all the occurrences of a definition that are encompassed by the definition.

(a) Using the definitions developed in Exercise 10, describe the width of each definition.

(b) For each definition, give an example of an occurrence that is covered and one that is not.

3

Data Item Sets: Top-Down Design

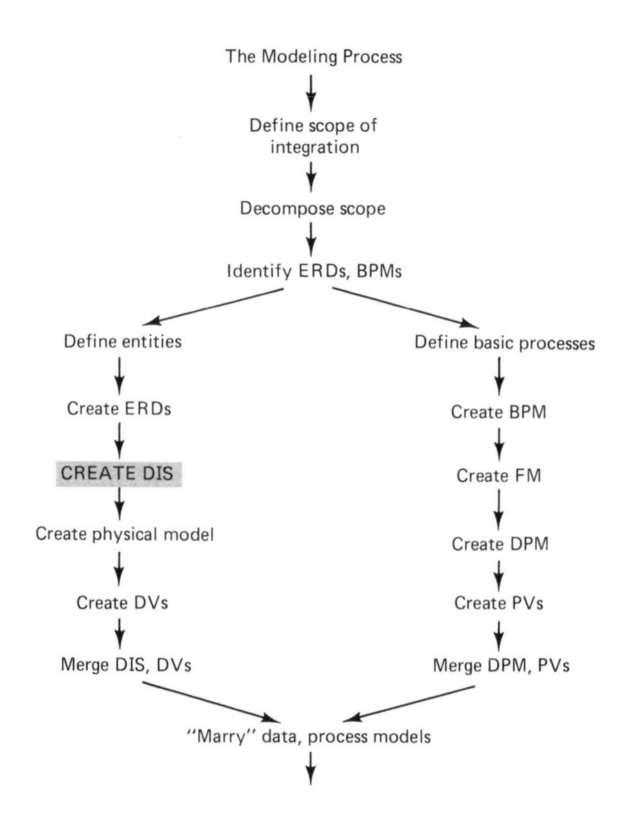

Once the entity-relationship diagram is complete with definitions rigorously defined, the next step in building the business-based system model is to extend the ERD. (*Note:* In parallel with this activity is an equivalent modeling activity for processes that will be discussed in later chapters.) Consider the banking ERD produced in the Project Study in Chapter 2, shown in Fig. 3.1. From this ERD one entity at a time will be expanded into a data item set. A data item set is merely an expansion of the entity in which a much greater level of detail is created. For the banking system, *all* entities would be expanded into their respective data item sets, but for the purposes of this book, because of space limitations, only one entity—account—will be expanded, as shown in Fig. 3.2.

The first step in translating an ERD entity into a data item set (or DIS) is to create a very broad outline of the components or attributes of the entity based on use of the data in the business environment. What goes into the DIS depends on how the data is perceived and used in the business of the enterprise. In this case there are five types of accounts: loans, savings, DDA, credit card, and trust accounts. In addition, there is data that is classified as an associate account (which was represented recursively at the ERD level). The first breakdown of acccount is shown in Fig. 3.3.

After an analysis of the type of accounts shown in Fig. 3.3, it is determined that a further breakdown is appropriate. A loan is broken into three kinds of loan—home, car, and personal—and there are three kinds of savings—passbook, Christmas club,

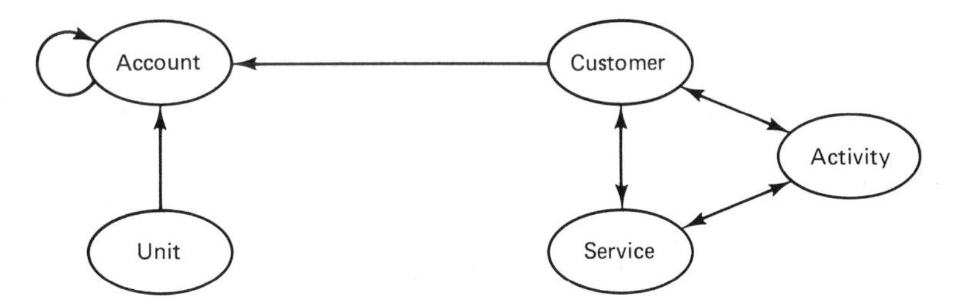

Figure 3.1 Banking ERD.

Figure 3.2 Account is expanded.

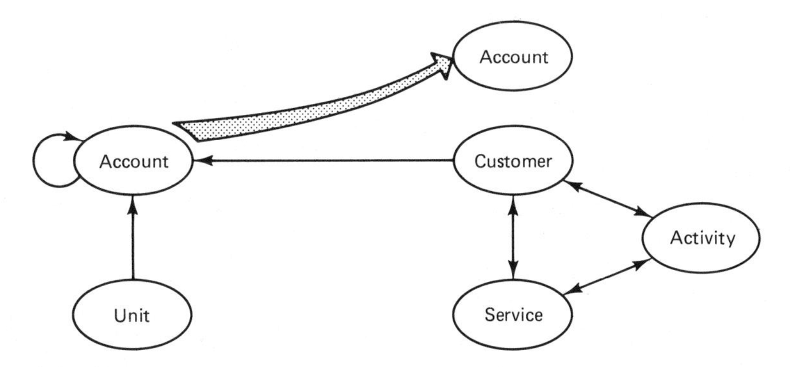

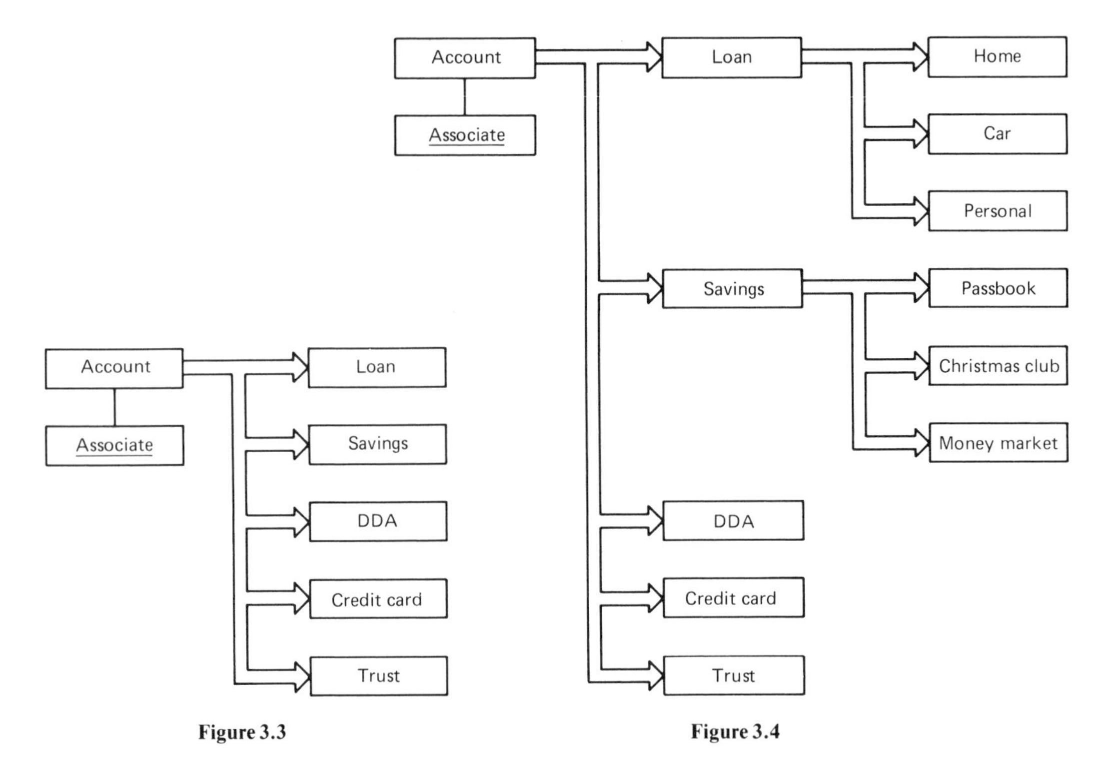

Figure 3.3

Figure 3.4

and money market. This breakdown makes use of the "is a type of" construct. There are no further classifications for DDA, credit card, or trust. The resulting structure is shown in Fig. 3.4.

Now that all major business aspects of an account have been represented, the first cut at the data item set is complete. It is now the task of the architect to begin to fill in the details. The first task is to complete the ERD relationship requirements. From the ERD there is a relationship between account and customer. This link is realized by means of an owner connector.

The use of the data in the business world is examined to determine what data elements are appropriate for each DIS. Next, each of the types of accounts are "fleshed out" in terms of the attributes that belong to them. For example, a loan has attributes of percentage, terms, type of loan, date due, and balloon payment. In addition, certain iterative data for a loan is recognized, such as payment data, penalty, due date, and interest paid. Attributes are determined from the processing that the data accommodate.

Other data attributes are determined for other types of accounts based on the business being transacted. For each account there is common activity that is recorded in a common place. Data recorded for common activity includes the amount, the form (cash, check, etc.), verification, location, and so on. The resulting data structure is shown by Fig. 3.5. To keep the structure as simple as possible, the attributes for home loan, car loan, personal loan, passbook savings, and so on are included in

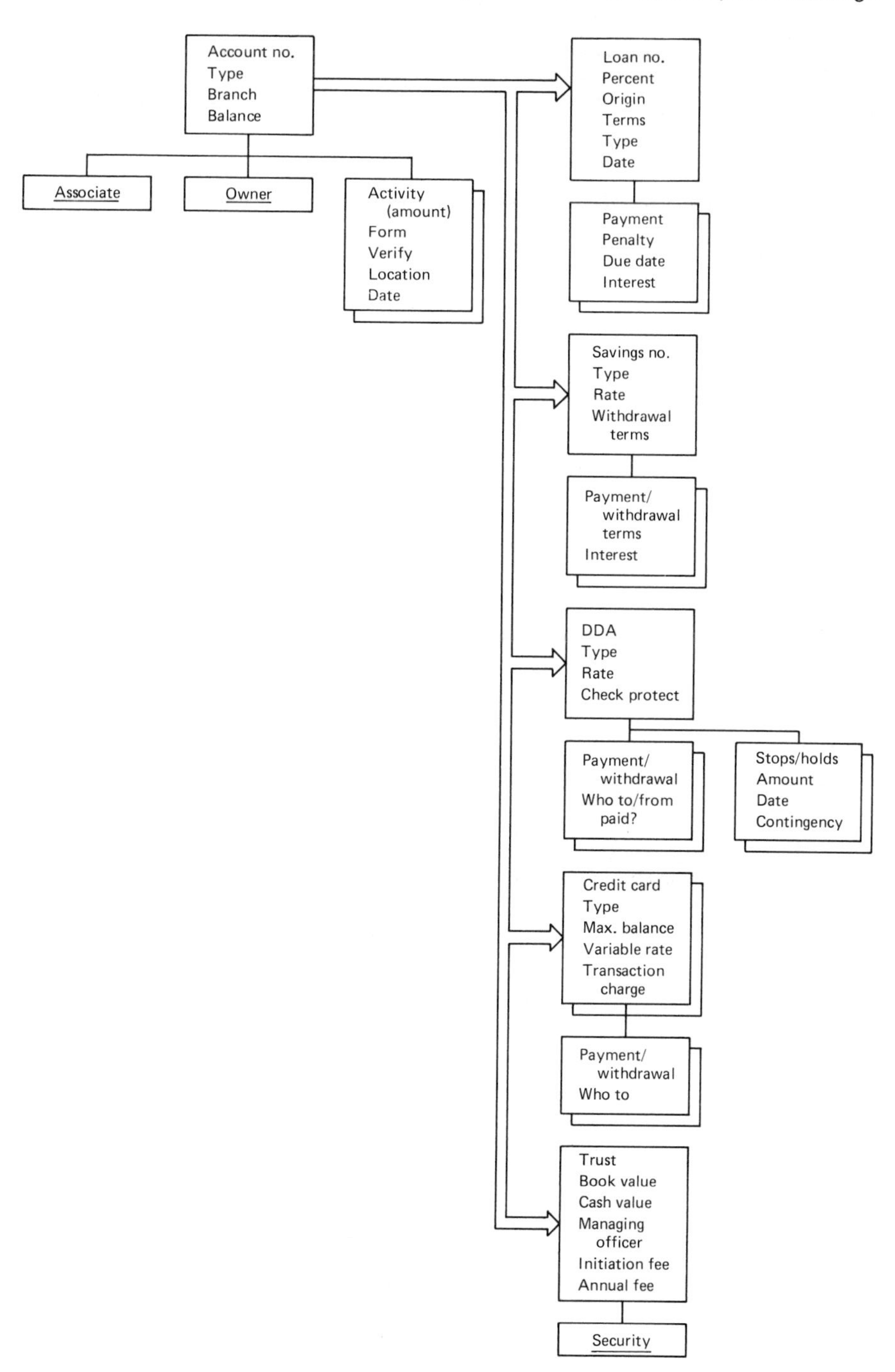

Account no.
Type
Branch
Balance
Associate
Owner
Activity
(amount)
Form
Verify
Location
Date
Loan no.
Percent
Origin
Terms
Type
Date
Payment
Penalty
Due date
Interest
Savings no.
Type
Rate
Withdrawal
terms
Payment/
withdrawal
terms
Interest
DDA
Type
Rate
Check protect
Payment/
withdrawal
Who to/from
paid?
Stops/holds
Amount
Date
Contingency
Credit card
Type
Max. balance
Variable rate
Transaction
charge
Payment/
withdrawal
Who to
Trust
Book value
Cash value
Managing
officer
Initiation fee
Annual fee
Security

Figure 3.5

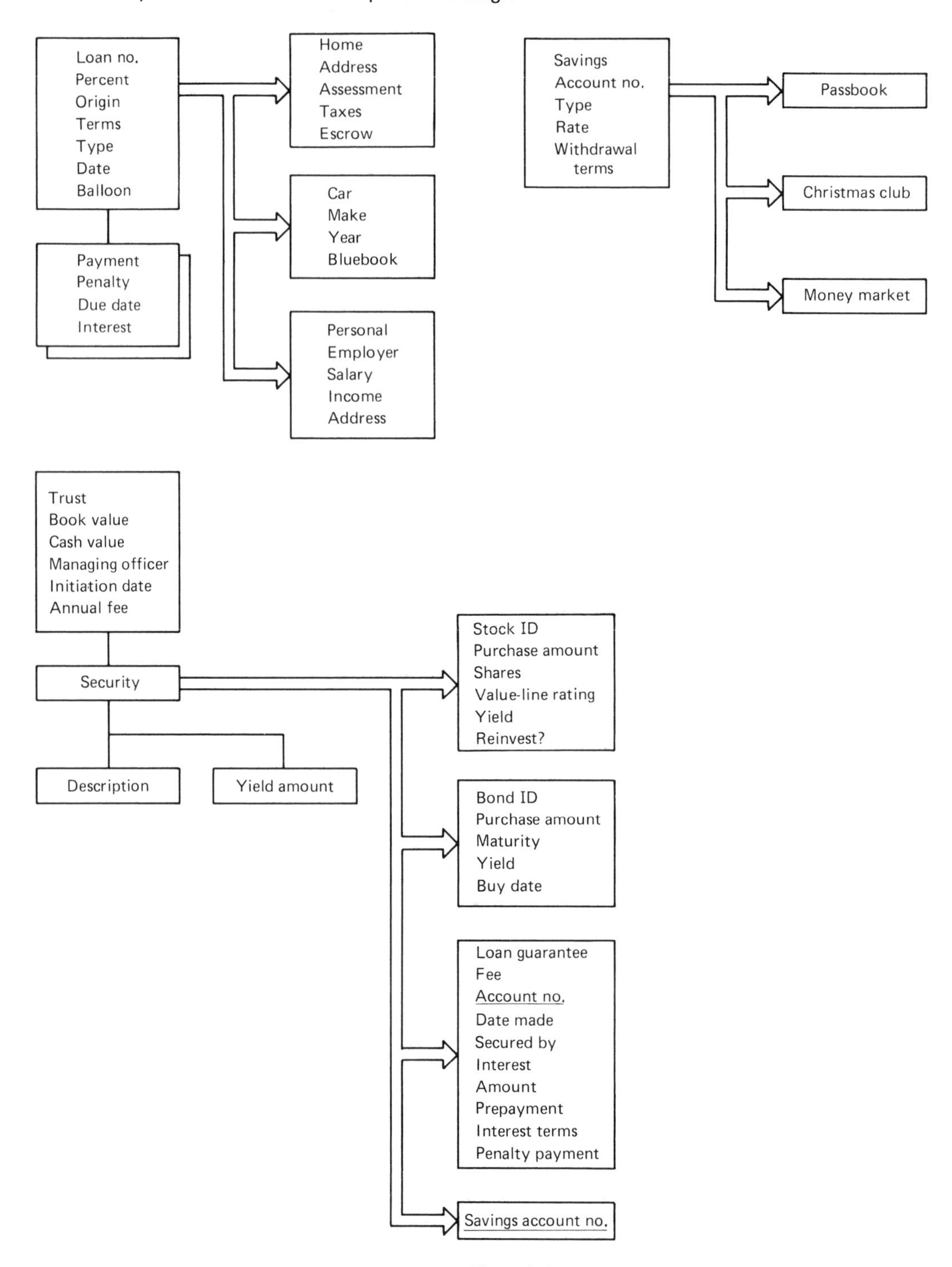
Loan no.
Percent
Origin
Terms
Type
Date
Balloon
Payment
Penalty
Due date
Interest
Home
Address
Assessment
Taxes
Escrow
Car
Make
Year
Bluebook
Personal
Employer
Salary
Income
Address
Savings
Account no.
Type
Rate
Withdrawal terms
Passbook
Christmas club
Money market
Trust
Book value
Cash value
Managing officer
Initiation date
Annual fee
Security
Description
Yield amount
Stock ID
Purchase amount
Shares
Value-line rating
Yield
Reinvest?
Bond ID
Purchase amount
Maturity
Yield
Buy date
Loan guarantee
Fee
Account no.
Date made
Secured by
Interest
Amount
Prepayment
Interest terms
Penalty payment
Savings account no.

Figure 3.6

separate documentation. Typical data items found for home loans are address, assessment, taxes, and escrow.

There are several types of security that exist as part of a trust account: stocks, bonds, loans, and savings accounts. There exists a description of security in another domain. The description is independent of whether or not the security exists beneath a trust account. In addition, from time to time a security yields a return, which is recorded as part of security information. This structuring of data (which is physically apart, but logically extended, from the structure shown in Fig 3.5) is depicted by Fig. 3.6, which represents a data item set prior to its "finished" form. There are some interesting points to be made about DIS construction in general, which will be illustrated in terms of the account DIS that has been developed.

DATA ITEM SET CONSTRUCTION

The first point of interest is an illustration of the concept of data existence dependency—the dependence of other data for existence. Dependency can be seen to exist on at least three levels:

- *ERD level:* One entity depends on another for existence.
- *DIS level:* A DIS exists only when the entity exists.
- *Sub-DIS level:* A data item exists only when a DIS exists.

As an example of dependency at the ERD level, an account does not exist unless a customer exists. At the DIS level, a loan payment does not exist if a loan does not exist. And at the sub-DIS level, interest payment on a loan does not exist if a loan payment does not exist.

The concept of dependency is deceptively simple until the possibilities for error are considered. Consider the structuring (which intuitively is very appealing) shown in Fig. 3.7. In this case owner and the owner's associated data *do not* depend on an

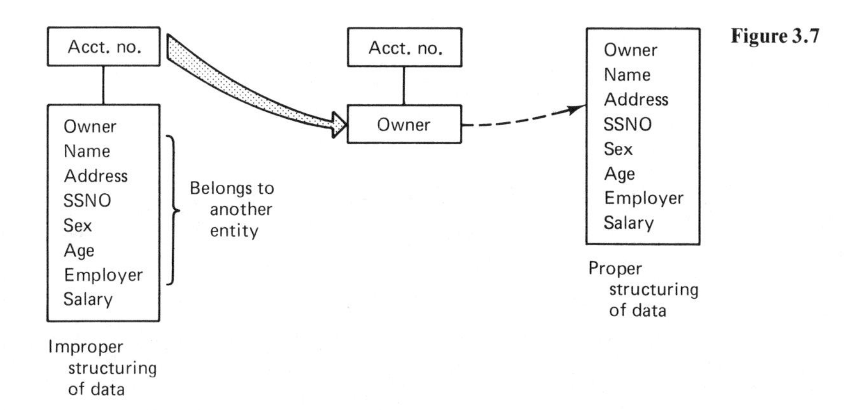

Figure 3.7

account for existence. They will exist independently of an account. Therefore, the proper structuring is to represent the relationship between owner and account number by means of a connector.

A second point of interest is the aggregation of data at the highest level. This is illustrated by branch and balance of an account. It is possible (but incorrect) for branch and balance of an account to be carried at the individual account level. But since branch and balance exist for *all* account types, they are carried at the account level, as shown in Fig. 3.8.

A third point of interest is in the notation being used to depict DIS. In one case, where iterations of data exist beneath a data item set, there is one and only one occurrence of the "parent" DIS. But in the case where a DIS is divided by an "is a type of" connection, there exists a parent DIS for *each* type of DIS. These relationships are denoted in Fig. 3.9.

A third point of interest concerning the notation used to structure DIS is that when a DIS is divided into types, the categorization must:

- Identify *all types.*
- Uniquely identify any given type.

This is shown in Fig. 3.10.

Figure 3.8

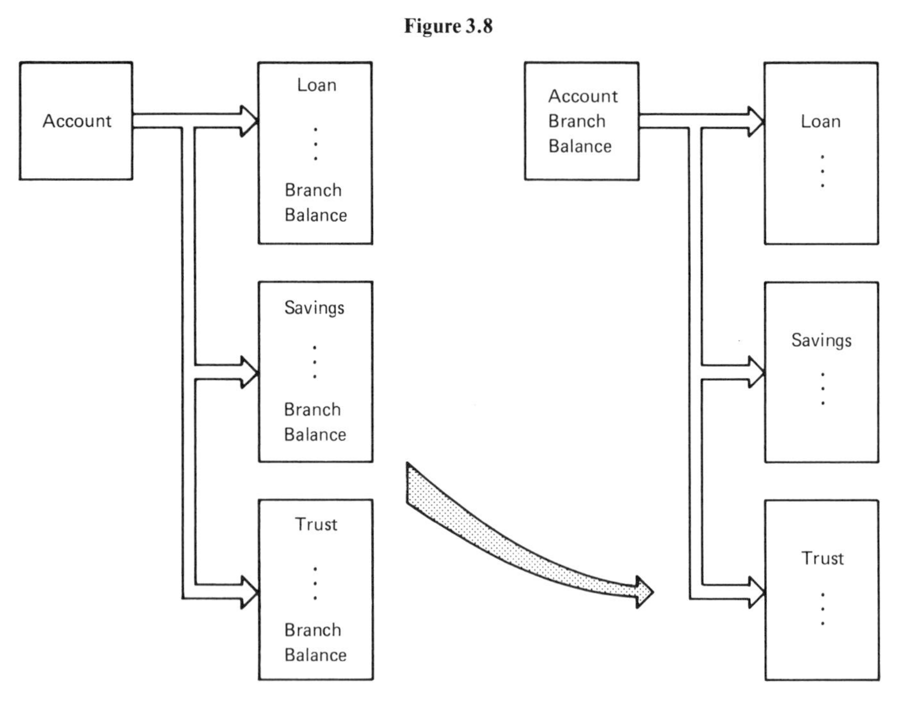

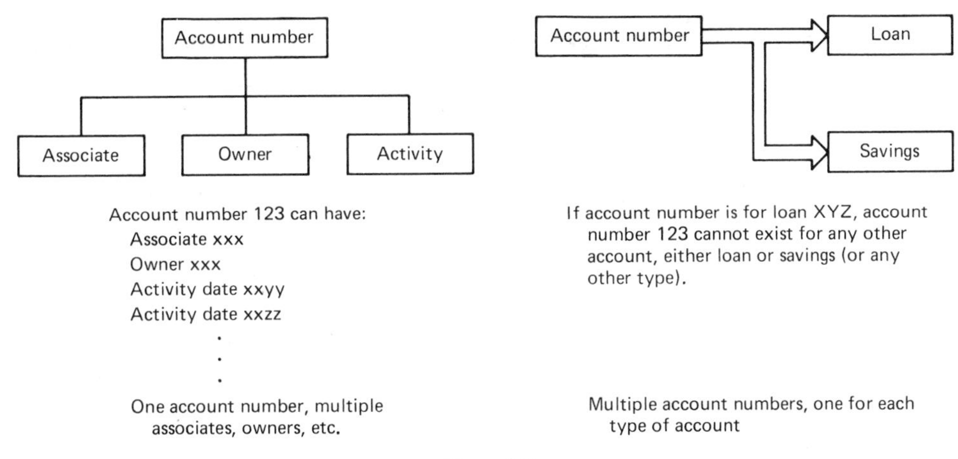

Figure 3.9

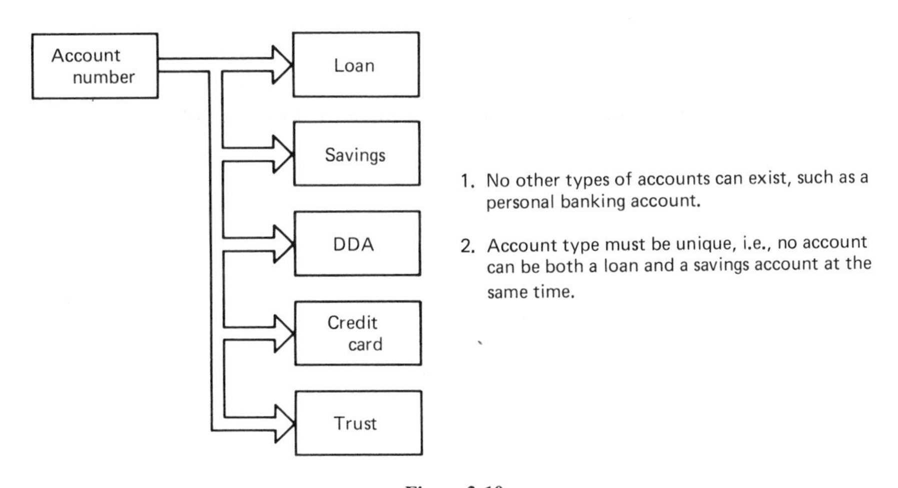

Figure 3.10

A point of major interest relates to the criteria by which a DIS is broken into types or categories. As a rule, the criteria that make the most fundamental or most pronounced division of the categories are the best criteria for typing or subdividing. (Note that it is normal to have *many* criteria for categorizing a DIS.) To illustrate the differences between different categorizing criteria, consider the following two ways to categorize a credit: by the individual types of credit card [Master Charge (MC), VISA, American Express (AMEX), etc.] or by individual or company credit card. The system being built is for a bank. Figure 3.11 illustrates the two choices being considered.

Which criterion is best for subdividing credit card: by type of credit card or whether the credit card is for an individual or a company? Both identify all types of

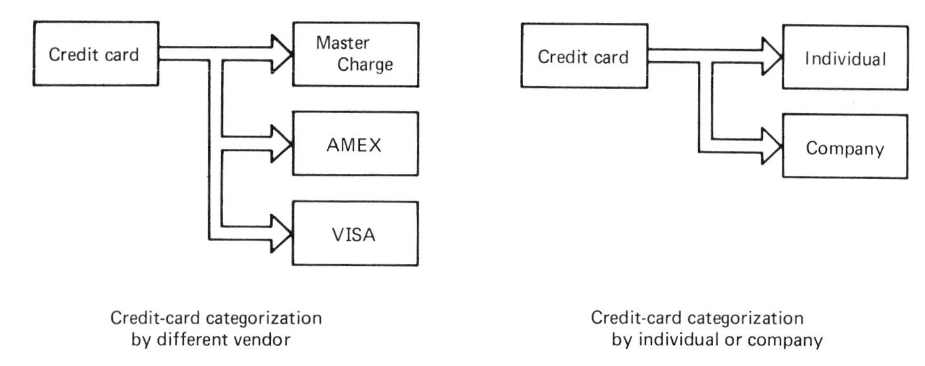

Figure 3.11 Two valid ways to type a credit card.

credit card (for a given bank) and both uniquely identify any given card. But division by vendor type is a better criterion because there are major differences between MC/AMEX/VISA as far as the bank is concerned, whereas there are only superficial differences between individual or company credit cards. MC/AMEX/VISA may calculate interest, monthly charges, and so on, differently, but there is little difference in card usage if the card is individualized or is for company use.

Another point of interest is that data from another dimension is seldom included in the construction of a DIS for a given dimension (of course, data from other dimensions can readily be acknowledged by a connector). The reason for seldom referencing data from other dimensions when constructing a DIS for a single dimension is that the data usually is farfetched (i.e., not central to the business being modeled). However, when the consolidated DIS is constructed, all dimensions will be accounted for. As an example, consider the data shown in Fig. 3.12.

A word is in order about the use of connectors in conjunction with DISs. Connectors exist in basically three forms: connectors to other dimensions, connectors to other DISs, and recursive connectors. These three types of connectors are shown in Fig. 3.13.

Figure 3.12

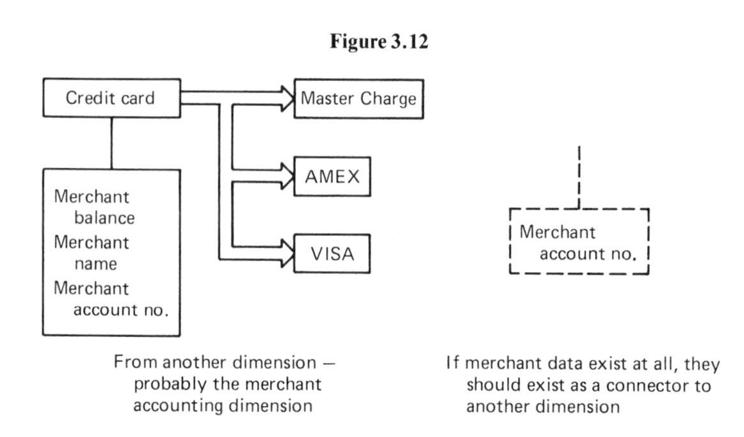

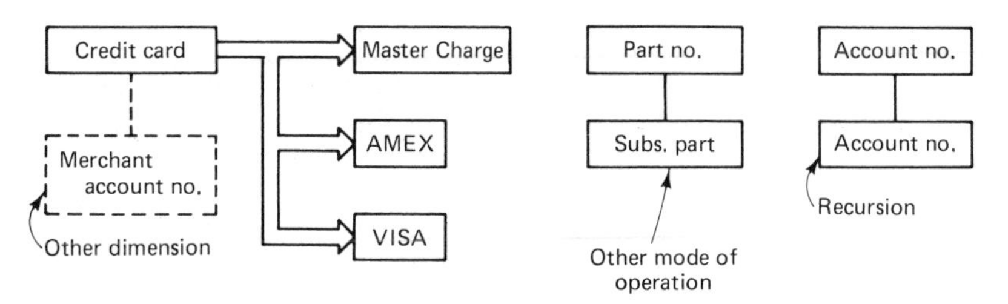

Figure 3.13 Three common uses of connectors.

A question faced by every system architect is: What elements should be included in a data item set? In general, any element necessary to the satisfaction of the function of the business dimension being modeled should be included. What should not be included are elements not necessary to the dimension. As a simple example of what *not* to include, consider Fig. 3.14.

Recursion is a commonly occurring phenomenon that bears special interest in representation at the DIS level. There is one generally proper form of representing recursion and it is through the use of a self-contained connector. It is normally *not* appropriate to represent recursion by iterating the different levels of the DIS. The proper and improper forms are shown in Fig. 3.15.

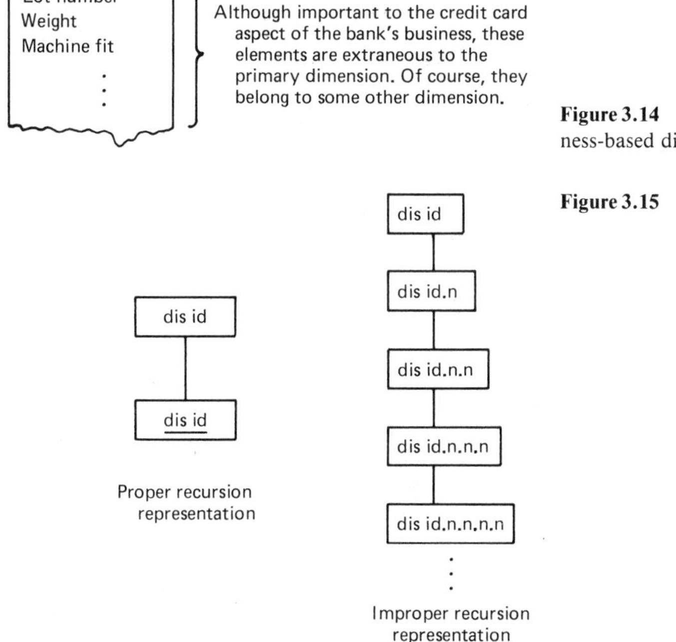

Figure 3.14 Dimension: primary business-based dimension of a bank.

Figure 3.15

There are several theoretical and practical reasons why there is a right way and a wrong way to represent recursion. In theory, a data item set (or any form of data) should be represented in its highest form. Since dis n, dis n.n, dis n.n.n, and so on, are *all* forms of a generic DIS, the proper representation is "dis." For example, in an automobile manufacturing environment dis n might be a car, dis n.n an engine, dis n.n.n a camshaft, and n.n.n.n a rotor. But a car, an engine, a camshaft, and a rotor are all forms of an assembled vehicle. From a modeling standpoint it makes sense to talk about different levels of vehicle assemblies rather than the actual individual assemblies. (This is another example of how the chain of abstraction applies to modeling. The higher level of abstraction—vehicle assembly—is more appropriate than the lower level of abstraction—the camshaft, the rotor, etc.) From a practical standpoint, when the many forms of DIS are represented separately, the associated data elements are also represented, thus resulting in redundant data definitions. This is shown in Fig. 3.16.

Another practical problem with a literally iterated representation of recursion is that there never is a convenient end to the recursive process. Still another problem is that often, units of the same generic level occur at different levels. For example, operating unit expenses normally occur as part of departmental expenses. But on occasion a division defines an operating unit immediately beneath it, which now puts operating unit at the plant level. Even more common is that an operating unit will have its own set of operating units. Now a literal representation becomes very complex and fragile.

As an example of recursion in the business world, consider the ordinary organization structure. For the purposes of the DIS the proper representation is as a managing unit having beneath it a reporting unit. A managing unit might be a divi-

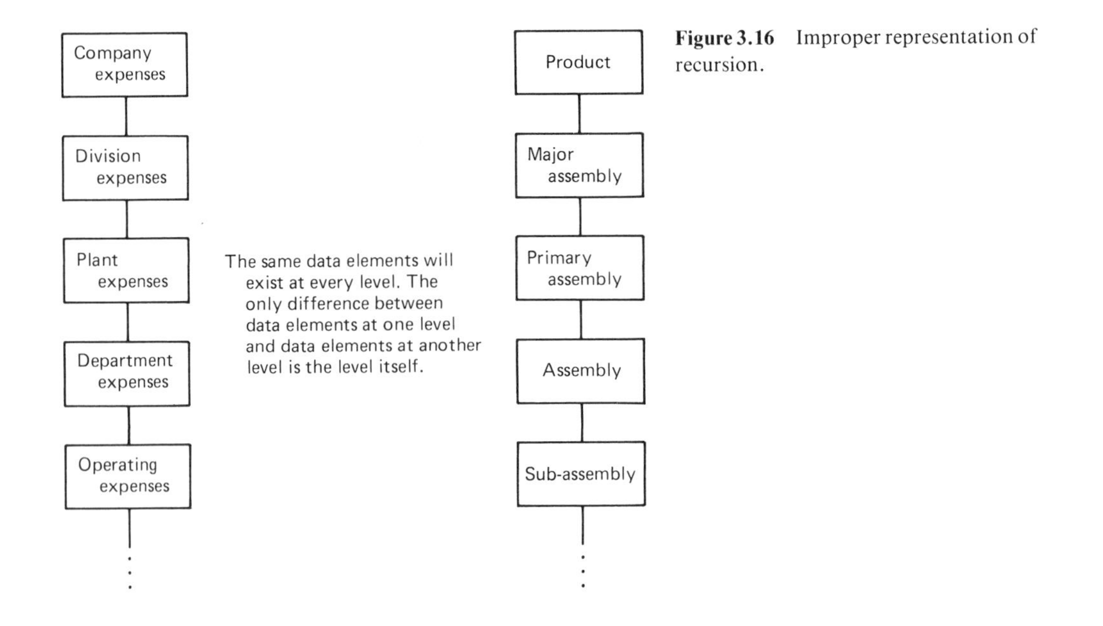

Figure 3.16 Improper representation of recursion.

sion and a reporting unit might be a department; in other cases a managing unit might be the department and the reporting unit a cost center. The improper structuring of recursion would look as follows: headquarters—division—department—cost center. There are several reasons why this structuring of recursion is a poor choice. What happens when there is a reporting unit below the cost center? What happens when a cost center reports directly to headquarters?

m:n RELATIONSHIPS

It is common to have *m:n* relationships at the ERD level. To be consistent, *m:n* relationships must be capable of being represented at the DIS level as well. However, there is no construct that conveniently lends itself to an *m:n* representation at the DIS level. But through a combination of existing constructs, the *m:n* relationship can be represented as shown in Fig. 3.17. (Note that the structures chosen lend themselves to any form of DBMS: hierarchical, network, or relational).

Another form of the *m:n* relationship occurs when an entity exists as a dependent of two or more entities. Such a relationship is shown in Fig. 3.18. In this case security exists as an intersection of account and customer. If either account or customer do not exist, the entity "security" cannot exist. The appropriate representation of security at the DIS level is shown in Fig. 3.19. Security exists as its own DIS, with connectors to both "account" and "customer." Note that data other than connector data can exist beneath "customer" and "account." Typically, this data would answer such questions as:

- Which account has associated with it a specific dollar value?
- What is the margin balance for a customer for a given security?

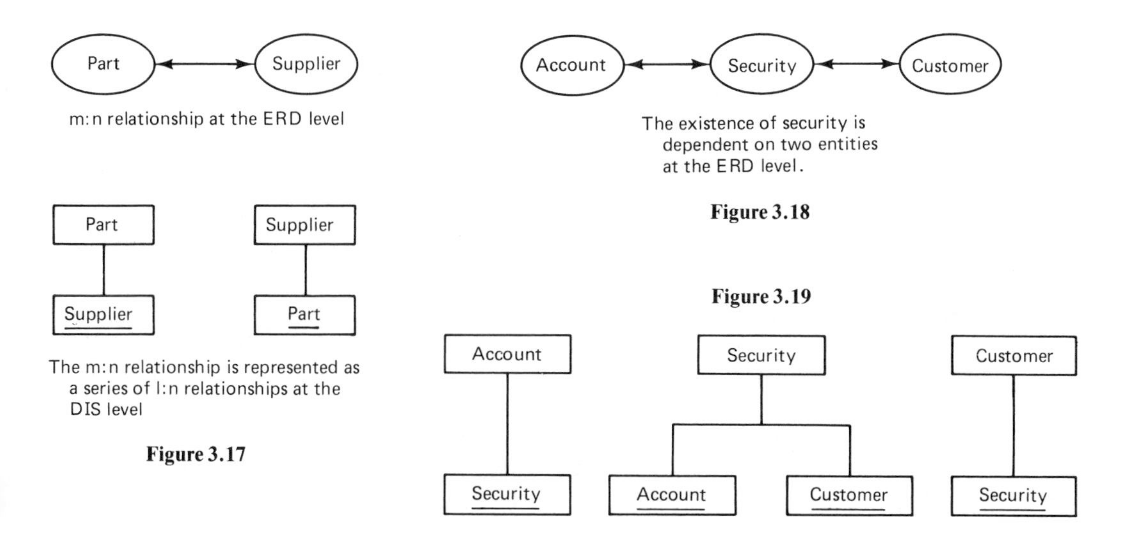

Figure 3.17

Figure 3.18

Figure 3.19

DATA ITEM SETS AND KEYS

One of the most important details with which the system architect must come to grips is that of the definition of keys within a DIS and within the types of DIS. Keys are the glue that binds together the many diverse types of data within the scope of integration. When keys have been defined properly, the system can expand for future needs, can satisfy current needs, and allows a user the means for transportation through the system. When keys are not defined properly, the system is less than satisfactory.

Suppose that a bank has identified a need for keys for three types of systems: savings, credit card, and loan. Each system has its own unique set of keys:

- *Savings:* 6 digits—alphanumeric
- *Credit card:* 21 digits—numeric
- *Loan:* 7 digits—alphanumeric

Unfortunately, the bank has not abstracted its data and does not realize that savings, credit card, and loans are all financial services and all serve the same thing—the customer.

Some of the difficulties that will predictably arise when there is a nonuniform keying of the same entity are:

- Savings programs do not recognize credit card activity.
- The customer must interface with three incompatible interfaces.
- Global customer treatment is nonuniform and very difficult, if not impossible, to accomplish.
- A single customer may have 3 IDs; logic must be written to connect the three accounts, and so on.

As the systems architect goes from the very high abstraction of the data model to the final form of implementation, it is best that there be a single coding scheme for all keys of the same type.

An alternative solution to the problems of key compatibility is translation of one key to another by means of an index. In this case a customer could determine his or her savings, credit card, and loan key by accessing an index. The customer enters the key that is known and a reference to an index produces the related key values. The advantage of this technique is uniform treatment of keys throughout all systems. The disadvantage comes in system performance, especially in light of significant on-line activity and/or a volatile data base, where customers are frequently changing accounts or frequently inquiring as to their other keys. Another disadvantage is an impairment of system availability. What happens when the index used for translation becomes unavailable (for a variety of reasons)?

A third alternative solution to key compatibility is algorithmic translation from one key to the next. This technique may be very simple or complex. It may involve something as simple as truncation of the digits of one key to produce the next or something as complex as hashing or randomization of keys. In determining the requirements for keys, the systems architect must look across all ERDs and across all phases of development to determine accurately the proper characteristics of a key. The perspective of the architect at this point must be as broad as possible, as the physical characteristics of the key must accommodate both current and future needs.

DATA ITEM SETS AND ENTITIES

The data item set represents an entity as it is viewed in a single dimension. It is normal for an entity to participate in many dimensions (although there will be only one primary business-based dimension). Thus a DIS, by itself, does not completely describe an entity. *All* the DISs that exist for the dimension are required to describe the entity *fully*. This relationship is shown in Fig. 3.20. The various DISs that describe an entity all belong to different dimensions. To describe the entity fully (and prepare for physical modeling), all the DISs must be consolidated. In doing so, a consistent definition at the key level can be formulated.

For example, suppose that the entity is a bank account. One dimension is that of customer and the DIS for that dimension would include loans, savings, and so on. For the vice-president of planning, the dimension that is appropriate includes the annual high balance, the annual low balance, activity frequency, and so on. For the bank auditor the dimension would include balance audits, security, cross-teller totals, and so on. Each of these dimensions legitimately affects the final consolidated DIS, even though not all belong to the primary business-based DIS. The key value selected for the consolidated DIS must accommodate *all* uses of the data.

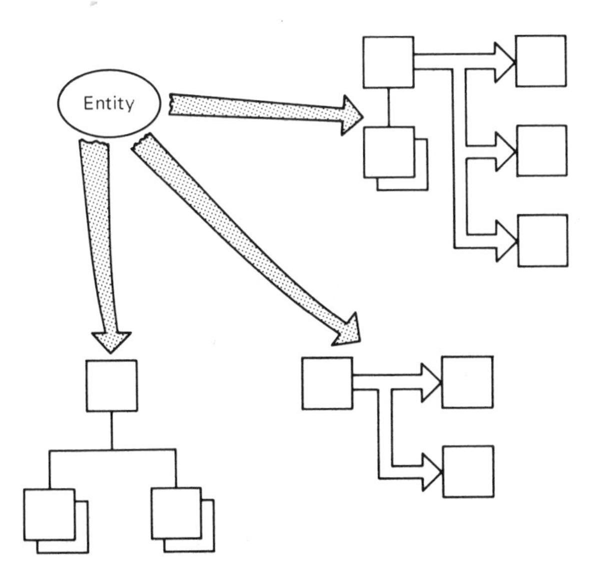

Figure 3.20 An entity is described by more than one DIS.

CONSOLIDATING THE DIS

Consider the following set of DISs for a part as they relate to order processing, bill of material processing, and inventory processing. Before the part can pass from the level of a DIS to physical design, its *total* requirements must be accounted for. This means that the various views must be consolidated. These DISs are shown in Fig. 3.21.

Figure 3.21 shows the DIS as it relates to the different types of processing. To be an effective model, it must incorporate *all* the requirements. This is done by nothing

Figure 3.21

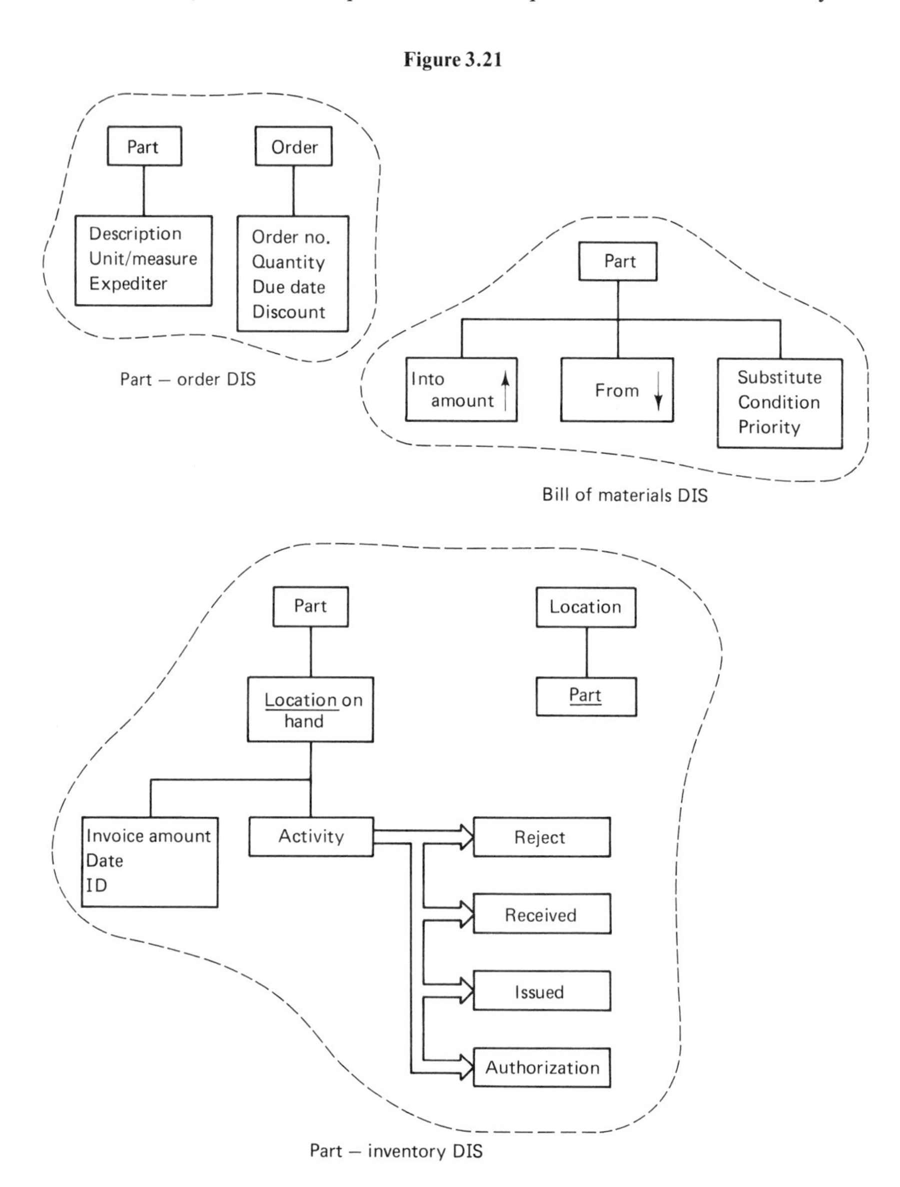

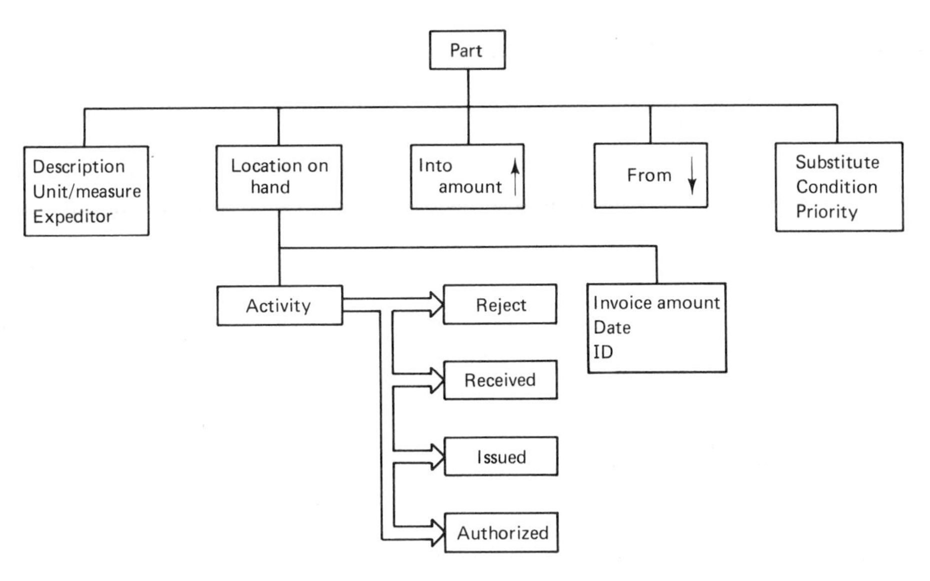

Figure 3.22 Consolidated DIS.

more than amalgamating the different relationships and data that exist for the part. Care must be taken so that two types of the same form do not end up with separate representations. Figure 3.22 illustrates an amalgamated DIS for a part based on the DIS of Fig. 3.21.

EXTERNAL PHYSICAL CHARACTERISTICS

After the consolidated DIS model is created, there is one final step that must be performed before the physical model is ready to be designed. That step is the identification of the external physical characteristics. The characteristics that should be specified (if appropriate) are:

- *Responsiveness:* on-line response time
- *Availability:* how often do the data need to be available? How long can the data be unavailable when it is unavailable?
- *Peak-period processing*
- *Volume of data:* initially, over time, growth
- *Environmental change:* where, with what probability?
- *Update responsibility:* which user, which elements?
- *Mode of interfaces:* direct, indirect, batch, on-line, etc.
- *Archival requirements*
- *Source of derivation*
- *Other data relationships/dependencies*

IN SUMMARY

The first step in building a business-based system model is the definition of the scope of integration. From the scope the business is decomposed, yielding the basic data and processes that form the backbone of the business. After the processes and data are defined for the primary business-based ERD, the entities of the ERD are expanded, as outlined in this chapter.

The DIS is a detailed extension of an entity. One feature of a DIS is that it is "categorized" (i.e., the various logical subdivisions of the DIS are outlined and detailed). Another feature is that detailed elements are included in the definition of a DIS so that the function that the DIS satisfies is able to be met.

DISs exist for different dimensions of data. To present a complete picture, all DISs that relate to an entity must be consolidated, which amounts to an amalgamation of the data necessary to satisfy the function for which the entity exists. One feature of the amalgamation process is the compatibility of the key across all dimensions in which the entity participates.

Finally, the external physical characteristics of all the DIS are described, in preparation for the building of the physical model of data.

PROJECT STUDY

After the scope of integration for ABC has been identified and an ERD constructed representing the primary business-based systems, the IA team decides to expand the ERD. Each entity is expanded into its own DIS. As an example of one such expansion, the entity "activity" is expanded into a DIS to include the element's date, amount, account identification, name, address, verification, and security identification. The physical characteristics of the DIS are determined. Of special importance are the physical characteristics of date, account identification, and name.

Date is of interest because it appears in several formats throughout the bank's many existing systems. Dates in the Julian calendar are selected as the most appropriate format since they can easily be translated to other formats in a standard fashion. Account identification is of interest because the physical dimensions must coincide or be compatible with several existing bank formats—loans, DDAs, and savings, to name a few.

The largest account identification key that ABC uses is a 16-byte DDA key. To allow for growth a 20-byte key is chosen so that all current ABC account keys can be accommodated. The IA team realizes that there will be wasted bytes of data, since accounts are the most commonly occurring data in the bank. However, the trade-off is made consciously.

The third field of interest is name. Name is to be a maximum of 25 bytes long. This will fit approximately 95% of ABC's customers. It is noteworthy that the appearance of a name on an activity is only for the purpose of identification of activity.

The name will not be used as a general-purpose cross-reference throughout the system. A name is used for clarification and auditing purposes only, and even then, in only 0.5% of the activities that occur. For the 5% of ABC's customers that will not fit into the 25-byte name limitation, there will first be truncation of the first name, followed by truncation of the lower-order characters of the last name.

Other characteristics of the data are specified, such as how many total activities there are for an account, how often the activities appear, how often the activities are accessed, the order in which the activities are accessed, and so on. The type of activity is considered, such as loan activity, savings activity, DDA activity, and so on. Not all data attributes will be necessary for each type of activity. In addition, loan activity requires two amount fields: payment and penalty. "Savings deposits" requires an identification as to source of payment and the amount of cash and noncash deposit. "Savings withdrawals" requires a field to be displayed with the current balance. "DDA" requires a field for system entry identification. The initial DIS for activity is laid out, as shown in Fig. 3.23.

After the DISs that describe the primary business-based ERD are laid out, other ERDs are built for the scope of integration. Two ERDs (representing different views of the data) are built: the audit ERD and the archived ERD. Although these two ERDs do not reflect the primary business-based view of the system, they do reflect legitimate business views.

It is found that the auditing of data requires an entry medium identification time and date stamp to be attached to each activity. The archival DIS brings up the fact that the loan balance must be registered after each payment. Another ERD (representing still another aspect of the date) that is built points out that account-to-account transfers require that two separate account numbers be specified. A DIS is built for these ERDs. Once built the DIS is consolidated with the primary business-based ERD. The resulting merger of DIS into the consolidated DIS for activity results in Fig. 3.24.

Figure 3.23

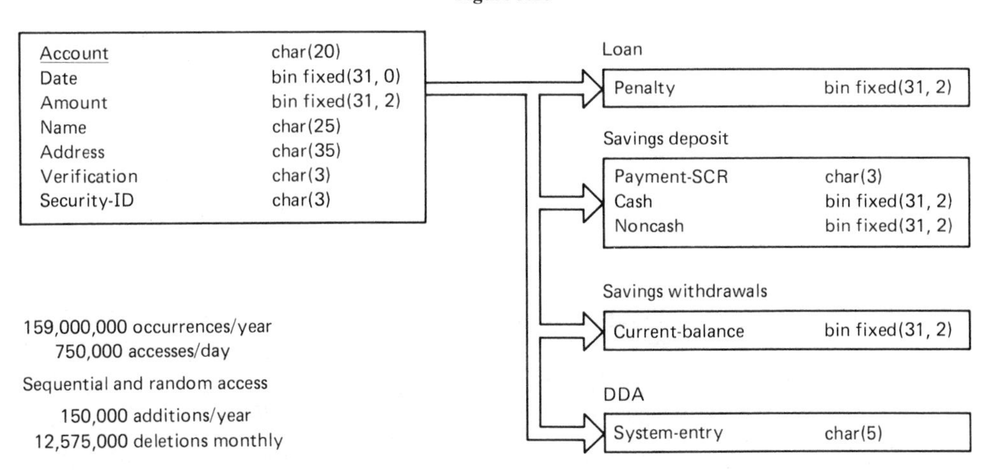

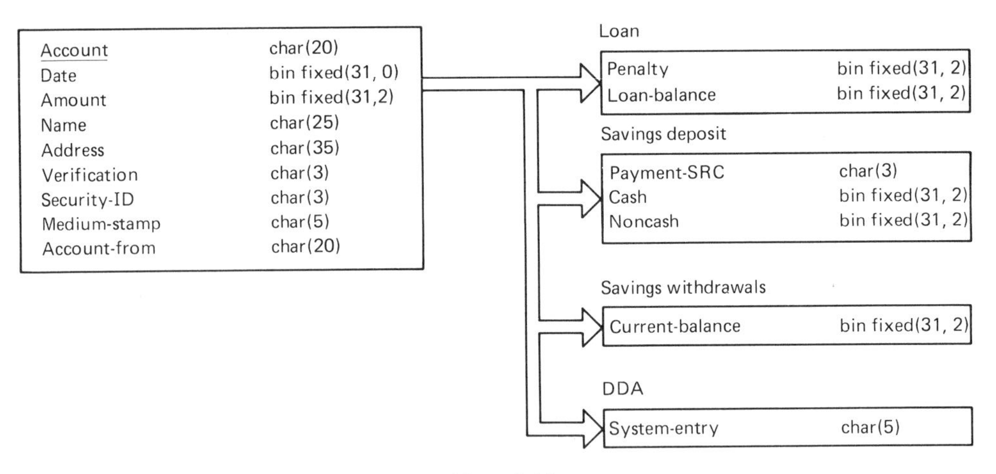

Figure 3.24

1. Why isn't the primary business-based ERD entirely adequate for the DIS specifications? Why is the primary business-based ERD the best place to start? What other DISs need to be aggregated? What DIS should not be included?
2. What are all the considerations in the selection of the physical criteria?
3. What are the consequences of choosing a data size that is too small? Too large? In the wrong format? Unable to accommodate existing formats? Unable to accommodate future formats?
4. How can optionally occurring data fields be handled? How can multiple occurring data fields be accommodated?
5. Is there ever a DIS with more than one unique key identifier? Why or why not?
6. Are there other "is a type of" categories for the activity DIS? If so, what are they?
7. What other DISs (other than activity) are there?
8. Draw the DIS structure for each of the DISs identified in question 7.
9. Identify all DISs that need to be built for the primary business-based ERD.
10. Identify several dimensions that should have ERDs created for them.

EXERCISES

1. A data item set (DIS) is an expansion of an entity in an ERD. Create a DIS for each of the following entities (which have been developed in the preceding chapters). Include attributes, keys, connectors, and all common types ("is a type of") of data.
   (a) Account
   (b) Part

(c) Supplier
(d) Customer
(e) Location
(f) Order
(g) Account activity

2. (a) What three types of connectors are commonly found? Give an example of each. Are these the only types of connectors there are?
   (b) How are connectors physically implemented? Can data be related without the use of connectors? If so, how?

3. (a) Give three examples of recursion.
   (b) Represent recursion in terms of a DIS format.
   (c) Why is a physical (i.e., literal) layout of data in its recursive form a bad idea? Is it ever justified? If so, when and under what circumstances?

4. Can an $m{:}n$ relationship be represented by a DIS? Should it be? What are the issues here?

5. (a) What is a key?
   (b) How can data be made unique?
   (c) Is uniqueness *beyond* the scope of integration relevant? Why not?
   (d) What is a generic key (or one used for alternative identification of data)?
   (e) Give examples of well-defined keys and of poorly defined keys. What makes one key definition good and another bad?

4

Physical Model: Top-Down Design

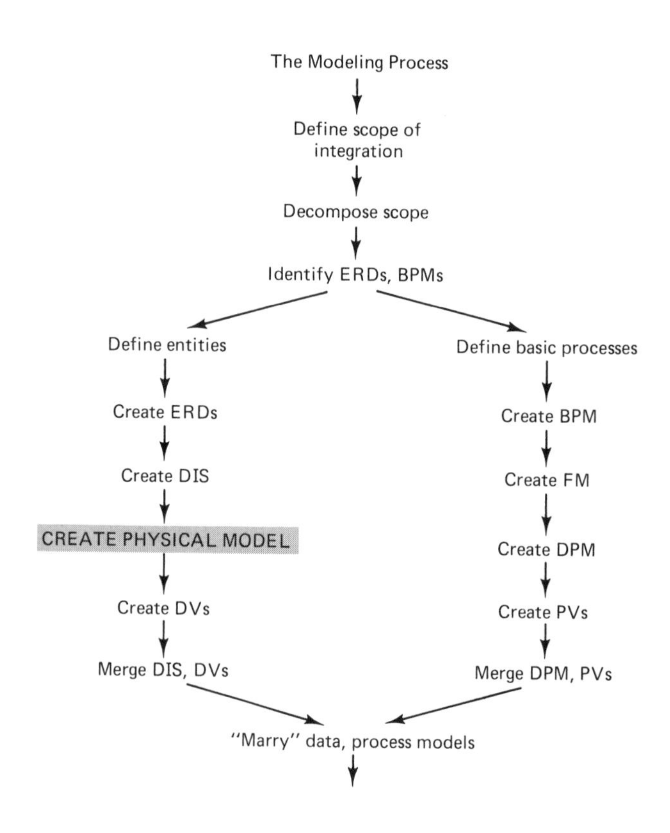

The first few steps going from the data item set to the physical model of data are strategic rather than functional. Some of the strategic considerations are:

- *Operational or decision-support systems:* Will the data be used primarily in one or in both environments?
- *On-line or batch:* Will the data be used exclusively, primarily, or evenly divided in the on-line or batch environments?
- *Large scale, midscale, small scale:* What sizes of data and processing are to be accommodated?
- *Hardware:* What are the hardware constraints under which the data must be modeled?
- *DBMS (and other software):* What are the software constraints for the data that is to be modeled?
- *Past experiences:* What are the experiences of the shop with regard to system successes and failures?
- *Criticality of data to the shop:* How close will the data be to the operation or management of the company?
- *Phases of integration:* What are the phases of integration? Which phase is currently being implemented? Which phases have already been implemented?
- *Existing systems:* What data currently exists in the shop? What data does not exist?
- *Political realities:* What are the political constraints on the design of the system?
- *Funding realities:* What are the limits of development with regard to what can be funded?

Each of these factors has an effect on the physical data model. Depending on the factor, the design may be simpler or more complex, performance oriented or flexibility oriented, a stripped-down version or a fully defined version, and so on. These factors must be considered *before* the model of the data can pass from conceptualization to a physical form.

WHY MODEL DATA?

There are many appealing reasons for modeling data and processes, but the single most important reason is to produce an integrated system. Of course, a high degree of integration significantly reduces system development time and system maintenance, thus improving development productivity. If it were not for the advantages of integration, there would be little real incentive to model data.

To achieve a high degree of integration, the physical model of data must "fit," or be compatible with, the preceding logical model of data or DIS, which of course fits with the preceding entity-relationship diagram. The relationship between the

ERD, DIS, and physical model is shown in Fig. 4.1. The physical model is based on the DIS and should fit with it. What exactly is meant by "fitting" with a DIS? Two criteria generally determine whether a fit is made:

- *Consistency of function:* Whatever function that is able to be accomplished throughout the scope of integration must be able to be accomplished uniformly within the different modes of operation within the scope. When the system is implemented, all functions within the scope of integration are able to be accomplished uniformly. For example, in a bank there needs to be consistency of certain kinds of functions, such as the close of business (monthly, annually, etc.). It does not make sense for one part of the bank to have one closing date and another part of the bank to have another.
- *Compatibility:* Data that is shared or that connects the same type of data must be compatible or have identical formats. The key structures of the systems within the same scope of integration must allow the user to pass easily from one system or mode of operation to another.

Figure 4.1

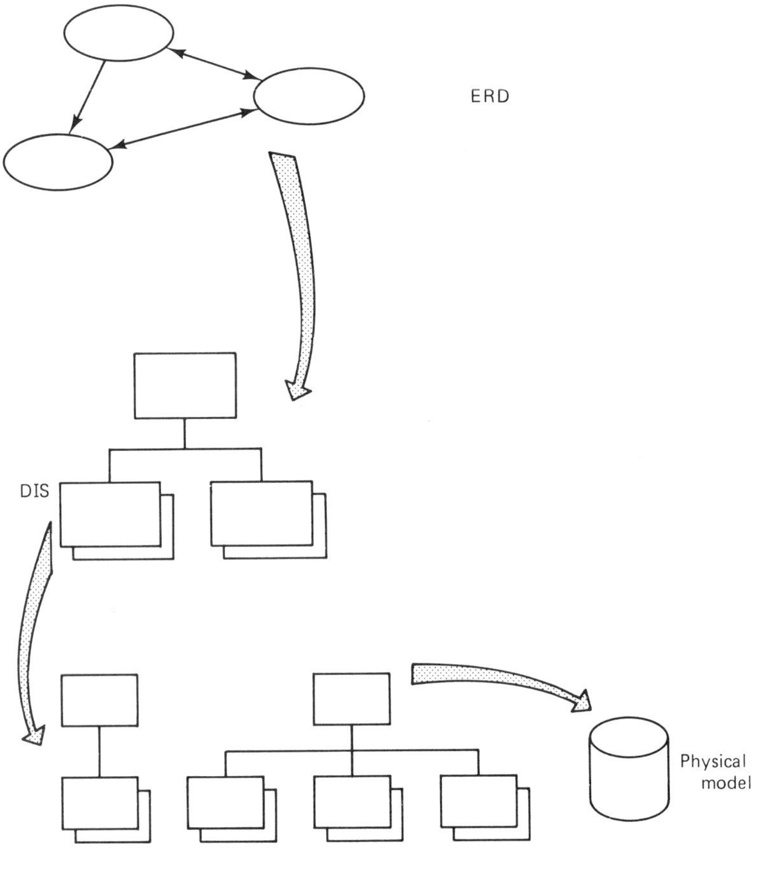

ERD

DIS

Physical model

As an example of each of these criteria, consider Fig. 4.2, which shows two physical models that have been designed for a part. The functionality that can be handled on the models includes a bill of materials capability, substitute part capability, and inventory capability. The data in physical model 1 is divided along functional lines (the bill of materials and substitute data are separated from the inventory data) as well as being divided by the type of data (raw goods are separated from assemblies). Note that the absence of a downward recursive pointer in the raw goods data base does not mean a loss of function, since raw goods do not come from any lower part. Also note that the data formats of the common data of raw goods and assembly are identical, but both formats are different from the inventory format. However, the *keys* connecting all three data bases are compatible. Thus the criterion of functionality is satis-

Figure 4.2

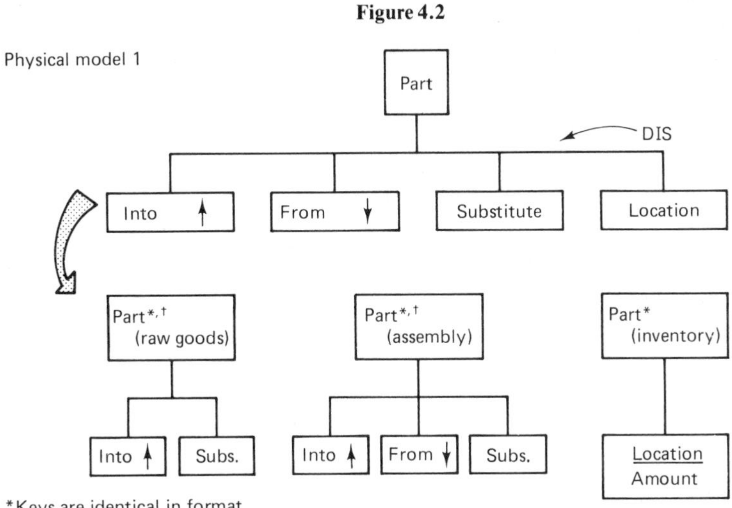

*Keys are identical in format.
† Data definitions are identical.

Contents of field definitions have the same meaning (e.g., when units of measure = lb — pounds for raw goods, then units of measure = lb — tons is not allowed).

Physical model 2

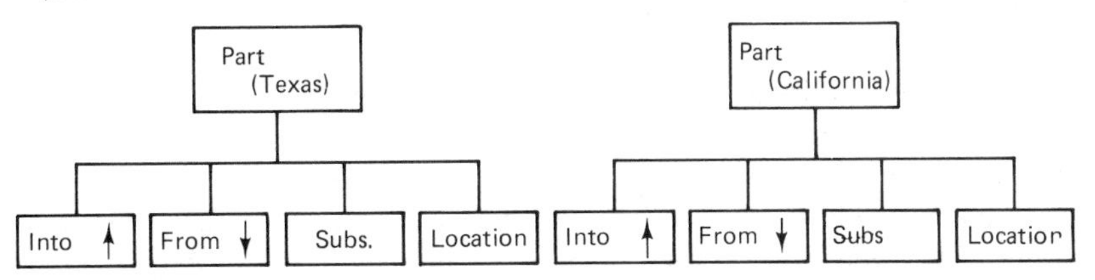

Keys are identical in format, but contents are separated by state.
Shared data definitions are identical in format.
Contents of field definitions have the same meaning.
Nonshared data definitions may be defined differently (e.g., Texas has a VAT tax for the manufacturing process but California does not).

fied by the collection of three physical data bases, while the criterion of compatibility is satisfied by a common key definition and layout compatibility of the raw goods and assembly data bases.

Another physical form of the DIS is shown by physical model 2. In this case data is split along state lines: Texas and California. Either data base fully satisfies the function required (for the state to which it is appropriate), and for data that is common to both, the format is identical. But in the example shown, Texas has a VAT (value-added tax) and California does not. Thus there is a difference between the data layouts insofar as noncommon data is concerned, but there is commonality elsewhere. The two criteria—consistency and compatibility—are thus satisfied.

As an example of an incorrect translation from the DIS to the physical model level, consider Fig. 4.3. Here savings account processing can be seen to be functionally incomplete in that associate accounts cannot be identified and the branch of the account does not exist. Thus the physical models *do not* satisfy the function shown at the DIS level.

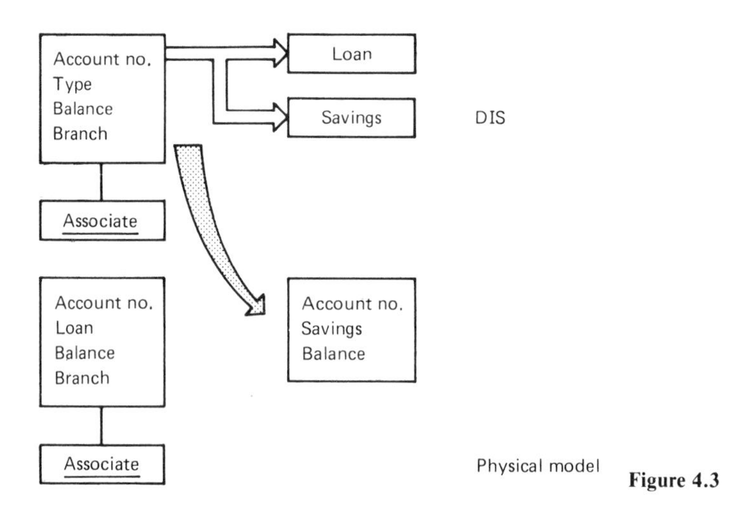

Figure 4.3

THE DATA ITEM SET MODEL: REQUIREMENTS

To translate adequately from the DIS to the physical model, the DIS must be defined *completely*. Because of the stringent relationship between the DIS and the physical model, the physical model can only be as complete and integrated as the DIS (since the DIS can only be as accurate as the ERD on which it is based). The following list of criteria, then, comprises what must be present in the DIS for the physical data model to be specified:

- The keys must be identified.
- Attributes must be identified.

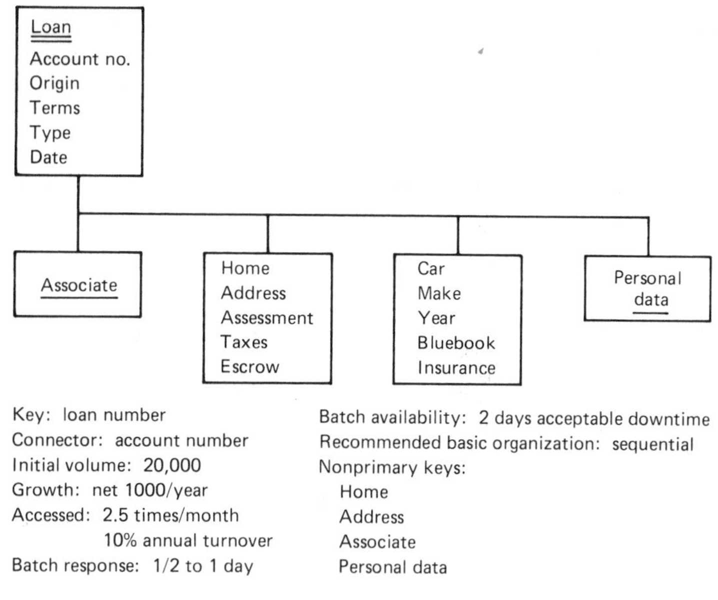

Figure 4.4

- The structuring of the DIS must be specified.
- External characteristics must be outlined.

Figure 4.4 illustrates what a physical model might look like. As the physical model is created from the DIS, several factors must be taken into account:

- The physical model entails a lower level of detail than the data item set.
- *All* dimensions are represented at the physical level, whereas at the higher levels, there is separation of dimensions.

REQUIREMENTS OF THE PHYSICAL MODEL

At first glance it appears that the translation from the DIS level to the physical level is straightforward, but in actuality there are some complicating factors in this translation. If *all* that was desired was a mere satisfaction of function, the translation would be straightforward. But the realities of the world (especially the on-line world) bring the factors of performance and availability into play. Both performance and availability must be designed into the system. They *cannot* be retrofitted. Tuning the system and adding hardware only marginally enhance the performance and availability of a poorly designed system. Thus the translation from the DIS to the physical level must take into account the needs of the data beyond the mere satisfaction of function.

The basis of performance is a careful use of I/O (input/output). I/O can be understood in terms of a trip that is made using bikes and Porsches. The trip begins in a Porsche, but soon the traveler must stop and ride a bike for several miles. Then the traveler gets into another Porsche and speeds off, until again, another bike must be ridden. It does not require much imagination to see that on such an odd trip, speed does not depend heavily on the speed of the Porsches, but on the number of times a bike must be ridden and the speed of the bikes. In the analogy the bike corresponds to performing an I/O operation, and the Porsche corresponds to performing a computer execution. The trip corresponds to the execution of an on-line program. A program executes very rapidly—at electronic speeds—until an I/O operation must be done, at which point the program operates at mechanical speeds. To achieve performance, then, the designer must be aware of I/O and design data physically so as to minimize the occurrences of I/O.

The second factor the designer must be aware of is availability. When an on-line system goes down, for any reason, it is unavailable to the user. The designer needs to be aware of two factors affecting availability: recovery and reorganization of data. Recovery occurs when data has a problem and must be backed up. Reorganization occurs when data needs to be restructured or "cleaned up" internally. For either recovery or reorganization, the limiting factor is the number of data that is physically defined together. The more data that is physically defined together, the longer recovery or reorganization takes and the longer the system is unavailable.

PHYSICAL DESIGN ALTERNATIVES

One of simplest ways that performance and availability can be enhanced at the physical design level is to subdivide or subset data. This technique is illustrated in Fig. 4.5, where the parts that go into a car are divided into five major classifications based on what general part of the car they are from. The key of each part is preceded by a letter describing the major classification, which is used to place the part in one of five distinct data bases. The structure of the data in the data bases is identical, allowing code that operates on one data base to operate on any data base.

Availability is enhanced in that when any one data base—say the engine data base—is unavailable for processing, the other data bases *are* available. Performance is enhanced in two ways. Because data bases can be processed separately and independently, it is easy to keep data internally highly organized so that the data can be processed efficiently. In addition, if the capacity of a processor begins to be exceeded, it is possible that one or more data bases can be removed to another processing environment, thus enhancing the overall performance profile. Note that this design technique *does not* violate the sound practice of integration by introducing redundancy. The structure of the data *is* redundant over the various data bases, but the data is nonredundant. No occurrence of data can exist in more than a single data base. And even though the structure is redundant, it is identical. There is a single definitive

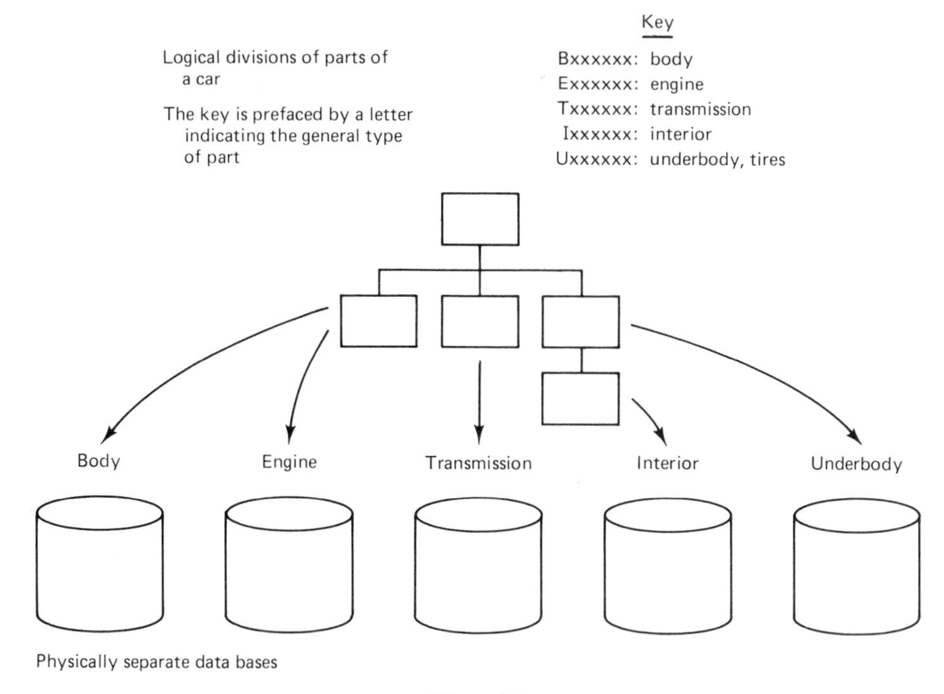

Figure 4.5

source for the structure. Were the structures not identical, the sound practices of integration would be violated.

Another technique for enhancing performance and availability at the physical design level (while still satisfying the criteria for integration) is shown in Fig. 4.6. The design techniques for performance and availability discussed *do not* violate the tenets of system integration because they do not require multiple source definitions of data or processes. Only one set of source code for data and processes is needed for an integrated system, even if the system is physically subdivided. The DIS model of data is shown being split into two physical data bases, A and A′. In data base A the divisions of data for A, E, and F and for B, C, and D have been combined, creating a two-tiered data structure. The two-tiered structure is a more efficient structure in which to store and retrieve data than is a multitiered structure, and requires less system management overhead. The gain in efficiency is somewhat offset by the loss of structural generality; that is, when C and D are stored physically independently of each other, a different number of occurrences of C and D can be accommodated, whereas when C and D are combined, only a fixed number of occurrences of C's and D's can be accommodated by the physical structure.

By splitting A and A′, both performance and availability are enhanced in that there is a minimization of storage problems. For example, when they are combined,

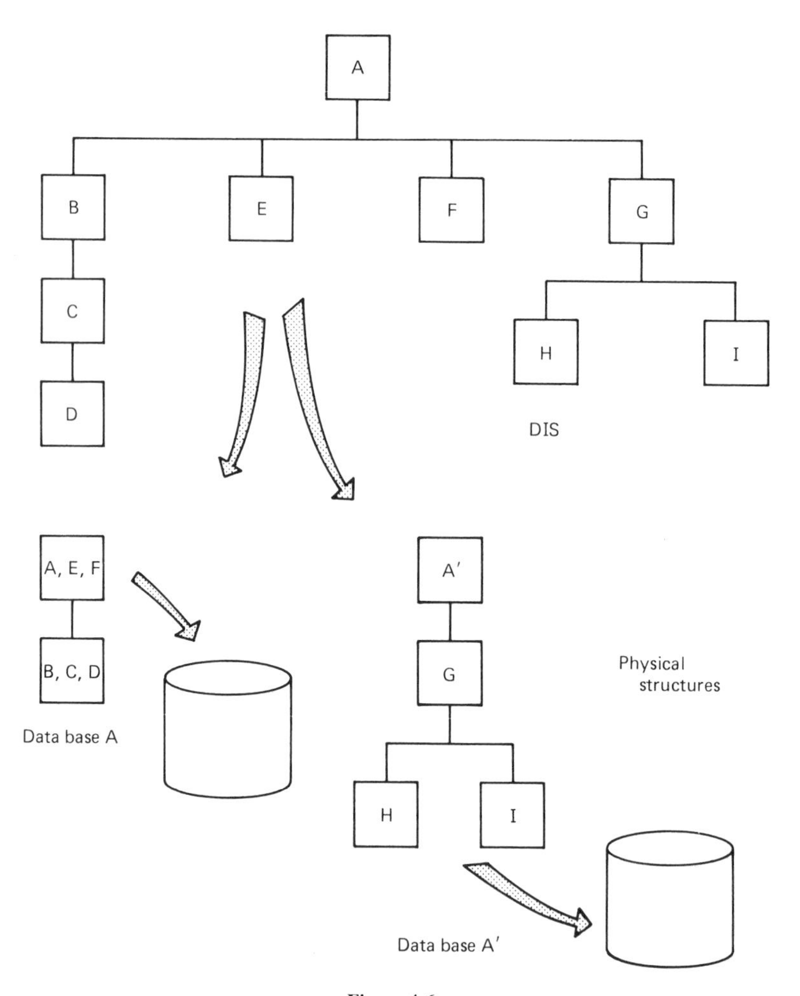

Figure 4.6

the physical layout of the data [as it exists on the (DASD) direct access storage device] tends to be awkward because of the mixing of the different types of data. The awkwardness occurs because the statistical variation of the bytes of data as they are stored is large. The more predictable the data is as stored on DASD, the more efficiently it can be stored. The predictability of the data in storage is more variable. The more types of data there are in a data base, the more unpredictable the data in storage becomes. This ultimate result is more I/O. Performance is then enhanced by separating types of data. (Availability is enhanced for the reasons discussed in Fig. 4.5.)

A word of caution: If A and A′ are to be split into physically separate data bases, care must be taken that there are not many processes that will need data from both A and A′ at the same time. If there are many such processes, the split will cause *extra* I/O (because they comprise physically separate data bases) and will thus *degrade* performance, not enhance it.

MEDIUM CONSIDERATIONS

Although most DIS transformation to the physical model is done for a data base (which resides on DASD), there are other media to which the DIS can be translated: principally tape, main memory, or photo-optical storage. Tape is used for cheap mass storage; main memory is used for limited, very fast access and storage of data; and photo-optical storage is used for large amounts of data that require little or no rewriting, such as might be found on an archival file.

Two techniques for transforming a DIS into a physical model for tape are shown in Fig. 4.7, where in one case each type of data—A, B, C, and D—is compressed into a single record DIS (i.e., a "flat file"), and in the other case, a separate record exists for each A, B, C, and D. Such techniques are referred to as a "flattening" of the DIS structure and are quite useful in building data structures that will be held on tape.

A typical DIS transformation of data into a format suitable for main memory is shown in Fig. 4.8. Here the DIS is stripped of its most important data and is presented as a table that resides in main memory. The cost of main memory and its many important uses other than for data storage require that an absolute minimum of data be kept, since main memory is a precious resource.

Figure 4.7 Translating a DIS into a "flat" physical model.

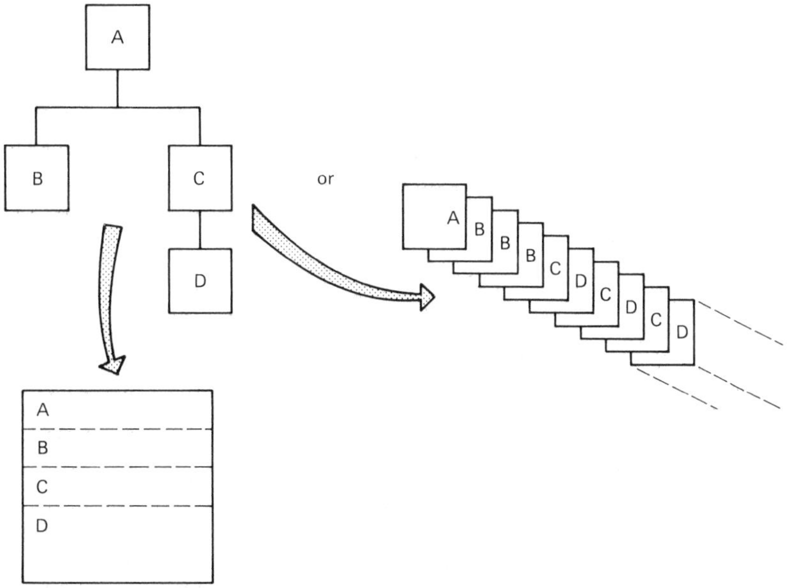

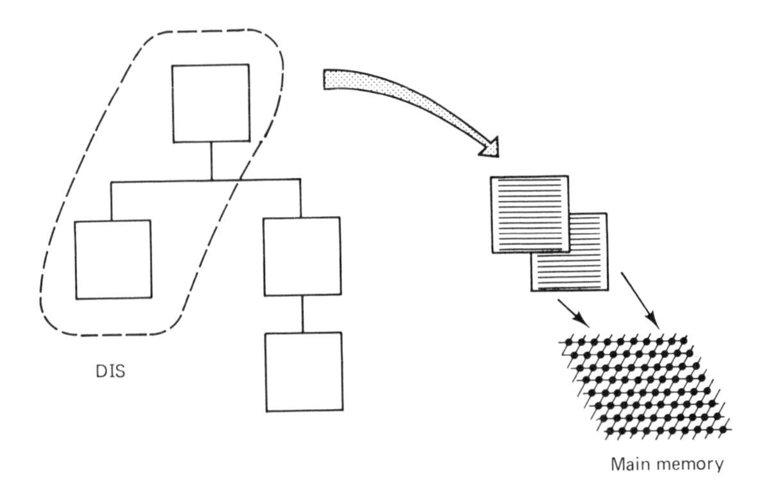

Figure 4.8

REPRESENTING CATEGORIES

The physical representation of categories of DIS is quite natural at the physical level and fits conveniently with most physical design practices. As a simple example of how a DIS that is broken into categories might be represented, consider Fig. 4.9, where the

Figure 4.9

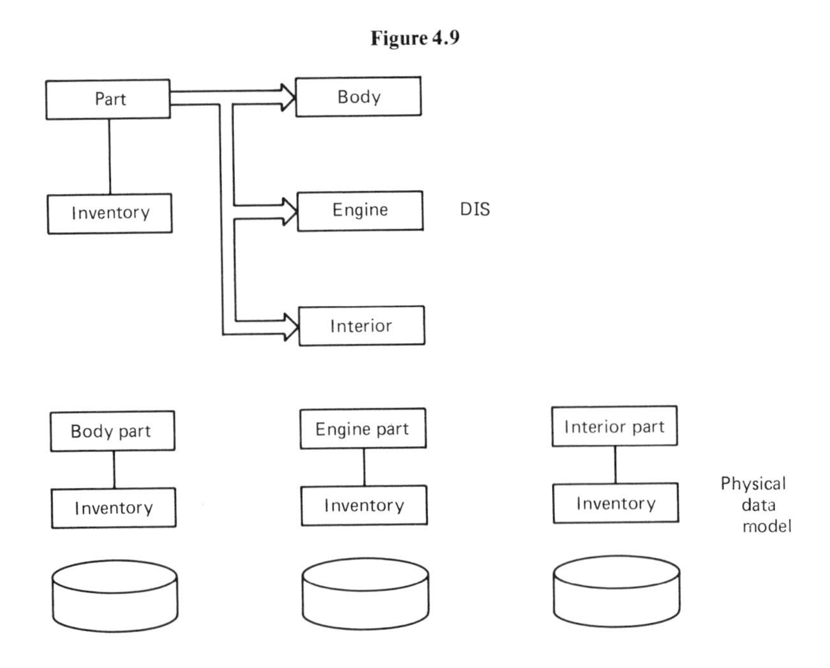

data has been divided according to categories of parts. For each part type the inventory is dependent on the part number. Note that the data and format are identical for all three structures (i.e., any code that operates on one data base is able to operate on the other data bases).

As a more complex example of translating a DIS into a physical model, consider Fig. 4.10. Here an account has "activity" and there are two kinds of accounts. The loans are divided into three types of loans: home, car, and personal. This organization is reflected by the DIS. The physical model is broken into four data bases. The data common to an account is found in savings and various types of loans. Any code that can operate on the common account data can operate on any of the four data bases. For savings there is unique code for data that is unique to savings. For each type of loan there is data and a corresponding code for data that is unique.

Figure 4.10

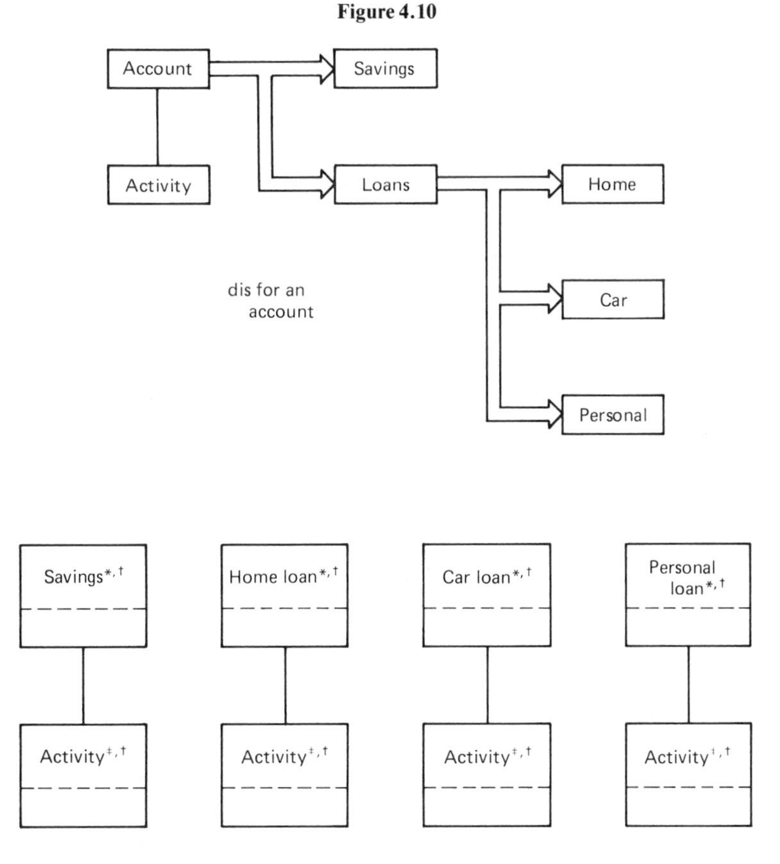

*Common format for account data.
† Format unique to type of account.
‡ Common format for activity data.

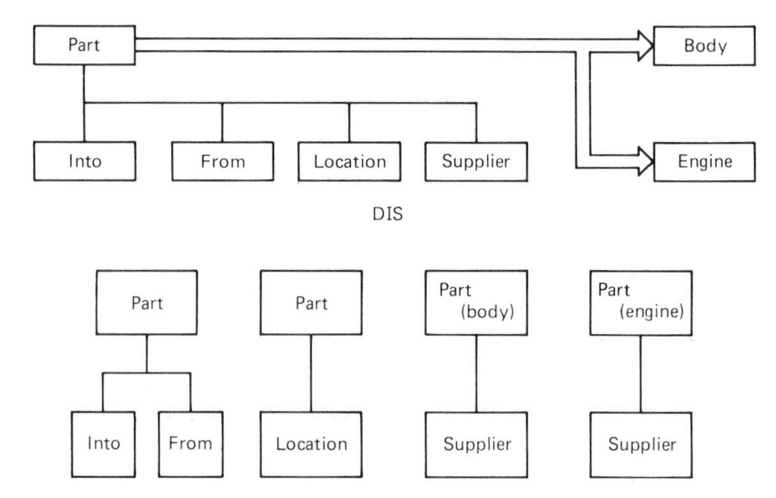

Figure 4.11 Physical data bases.

For a third variation on representing DIS physically, refer to Fig. 4.11, where there are four physical data bases for the DIS. The key is common throughout the four data bases. The data base representing the bill of materials (with the "from" and "into" connectors) contains *all* the data in the data base (i.e., all bill of materials data for body and engine). Similarly, the inventory data bases (with the location connector) contain both body and engine data. But the format of the bill of materials and the inventory data bases is different—only the key connects them. The other two data bases are the supplier data bases for body and engine parts. The formats of these two data bases are identical to each other but different from the bill of materials and inventory data bases. Collectively, these data bases fulfill the function represented in the model DIS.

OTHER PHYSICAL REPRESENTATIONS

As a shop goes to a physical model of data, there are strategic considerations as well as purely functional considerations. The physical model of the data is profoundly affected by the mode of usage of the data. Consider, for example, the operational, archival, and decision-support modes, depicted in Fig. 4.12. In this figure, archival systems are seen to be simplified and stripped down, operational systems are seen to be very similar or identical to the DIS, and decision-support systems are supported by a proliferation of the basic form of the DIS. The different modes are held together by a common key structure and a common understanding of the data as presented by the DIS.

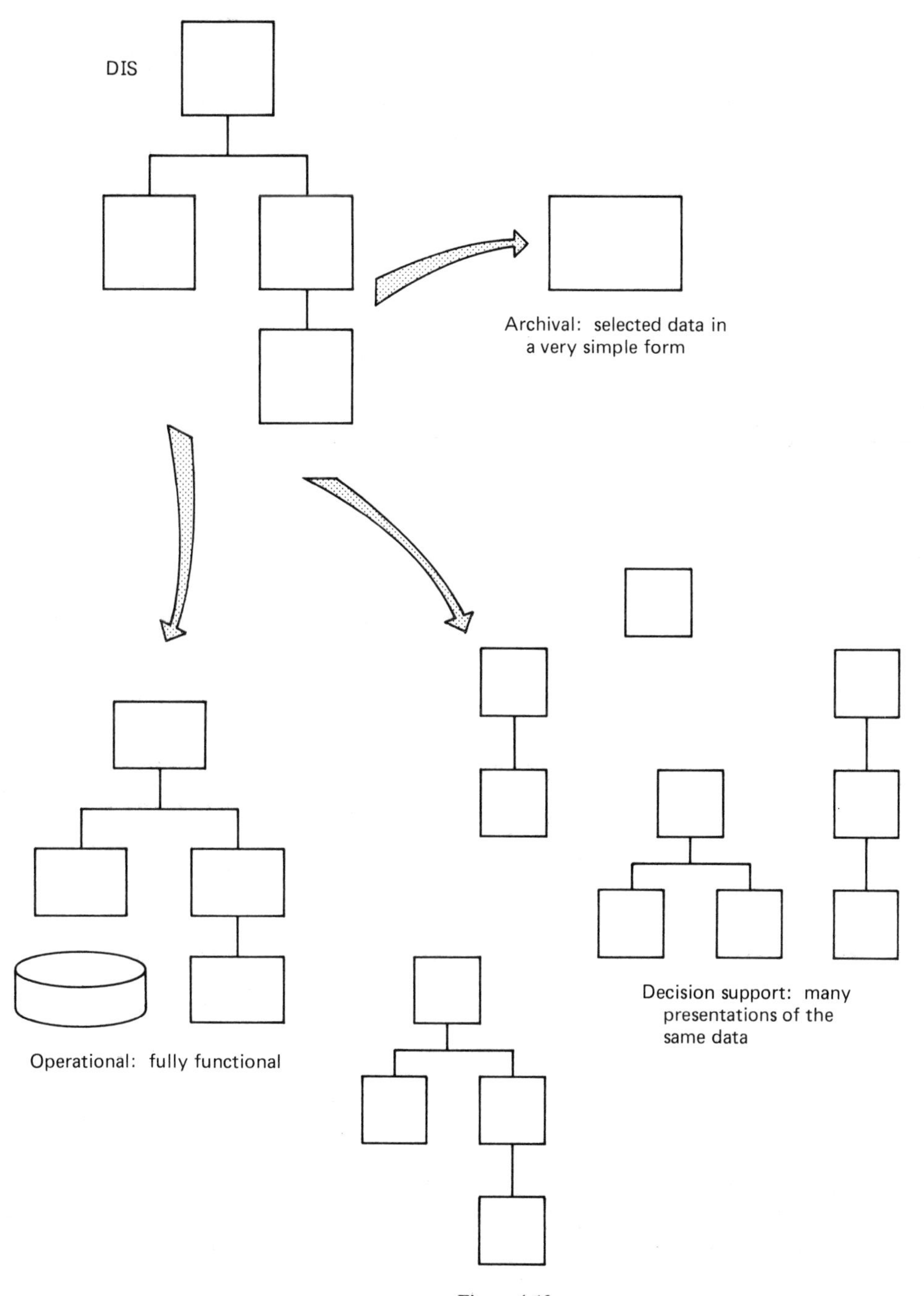
DIS
Archival: selected data in
a very simple form
Operational: fully functional
Decision support: many
presentations of the
same data

Figure 4.12

COMPLETING THE PHYSICAL MODEL

After the DIS has been translated into an appropriate physical model, there remain some basic decisions to be made, such as:

- Access methods
- Connectors
- Blocking
- Field lengths
- Alternate access routes
- Subsetting of data
- Key specifications
- Other detailed decisions

Note that there has been a radical shift in responsibilities in the design of a business-based, modeled, integrated environment. In traditional design, the application designer makes most decisions based on an immediate, local view of the user's requirements. In an integrated environment most of the decisions are already made by the time the design reaches the application designer. The result of the shift in responsibilities is a much more stable system that takes into account a much wider perspective.

IN SUMMARY

The first step in building a business-based model is a determination of the scope of integration, followed by ERD definition and process definition. Then the DIS is constructed from each entity in the ERD. Once all dimensions are accounted for, the DIS is ready to be consolidated and the physical model is ready to be built.

The first set of considerations in translating the DIS to a physical model are strategic in nature. The primary strategic factors are the operational and decision-support environment, batch or on-line environment, scope of system size, and so on. The criteria of whether the translated physical model fits with the DIS are:

- Consistency
- Compatibility

Physical design factors that will greatly influence the design are performance and availability of the data. Several techniques that are useful here are the splitting or combining of data where appropriate. There are several standard techniques to use in representing types of DISs, all of which are consistent with the goals of performance, availability, and systems integration. Finally, as a shop shifts its design practices from traditional piecemeal design to integrated design, many traditional design decisions are removed from the authority of the application designer in the interest of serving a broader (or global) perspective.

PROJECT STUDY

The creation of the physical model of ABC from the DIS is a straightforward process as far as the content of the data is concerned. The activity data look as described in Fig. 4.13. The primary considerations as the data is translated from the DIS to the physical model are (1) satisfaction of function, (2) performance, and (3) availability. The layout of the activity data is variable in length depending on the type of activity represented. Certain fields occur in every record and certain fields occur only in records of a certain type.

After the contents of the record layout are determined there are other physical design criteria. For the purpose of availability, the data bases containing activity will carry only limited amounts of activity. This will limit the sizes of the data bases, making them more easily recoverable, which in turn enhances availability. DDA activity (checking account activity) will go back one month (DDA activity makes up 75% of all activity handled by ABC). Loan activity will be stored for six months, as will savings activity. All other activity will be archived. When data is archived, it is

Figure 4.13

Account	char(20)
Account-type	char(1)
Date	bin fixed(31, 0)
Amount	bin fixed(31, 2)
Name	char(25)
Address	char(35)
Verification	char(3)
Security-ID	char(3)
Medium-stamp	char(5)
Account-from	char(20)
Penalty	bin fixed(31, 2)
Loan-balance	bin fixed(31, 2)

Variable-length record layout for loan activity data

Account	char(20)
Account-type	char(1)
Date	bin fixed(31, 0)
Amount	bin fixed(31, 2)
Name	char(25)
Address	char(35)
Verification	char(3)
Security-ID	char(3)
Medium-stamp	char(5)
Account-from	char(20)
Payment-SRC	char(3)
Cash	bin fixed(31, 2)
Noncash	bin fixed(31, 2)

Variable-length record layout for savings deposit activity

Account	char(20)
Account-type	char(1)
Date	bin fixed(31, 0)
Amount	bin fixed(31, 2)
Name	char(25)
Address	char(35)
Verification	char(3)
Security-ID	char(3)
Medium-stamp	char(5)
Account-from	char(20)
Current-balance	bin fixed(31, 2)

Variable-length record layout for savings withdrawal activity

Account	char(20)
Account-type	char(1)
Date	bin fixed(31, 0)
Amount	bin fixed(31, 2)
Name	char(25)
Address	char(35)
Verification	char(3)
Security-ID	char(3)
Medium-stamp	char(5)
Account-from	char(20)
System-entry	char(5)

Variable-length record layout for dda activity

still available to the user but on an ad hoc basis, with at least a day's turnaround time, while the massive amount of archived data can be searched.

Activity will be stored in reverse date order since inquiries against activity have the highest chance of looking at the most recent data first. Activities will be added in bulk at night, or a teller can manually enter an activity at the request of a customer. Direct activity entry occurs in about 0.017% of all cases.

For commercial accounts with a high amount of activity (account deposits, etc.) separate data bases will be kept (i.e., low-volume accounts will be on one data base and high-volume accounts will be on another). The separation of data bases will allow voluminous accounts to be removed off-line to be searched sequentially, thereby enhancing the performance of the on-line system. No secondary indexes will be established, although the IA team notes that it may become necessary later to establish them on fields such as date and amount.

(*Note:* So far, only the expansion of the ERD entity "activity" has been shown in detail. But concurrent to this activity is the fleshing out of the other entities: customer, service agreement, and account. The fleshing out of the other entities has not been included in the book because of space limitations.)

The physical data base design that has resulted from the DIS expansion is shown in Fig. 4.14. Seven physical data bases have been specified. They are: a loans data base, a savings data base, a DDA data base, a bankcard data base, a customer data base, an archival data base, and a service agreement data base. In addition, once

Figure 4.14 ABC's physical data bases based on the DIS.

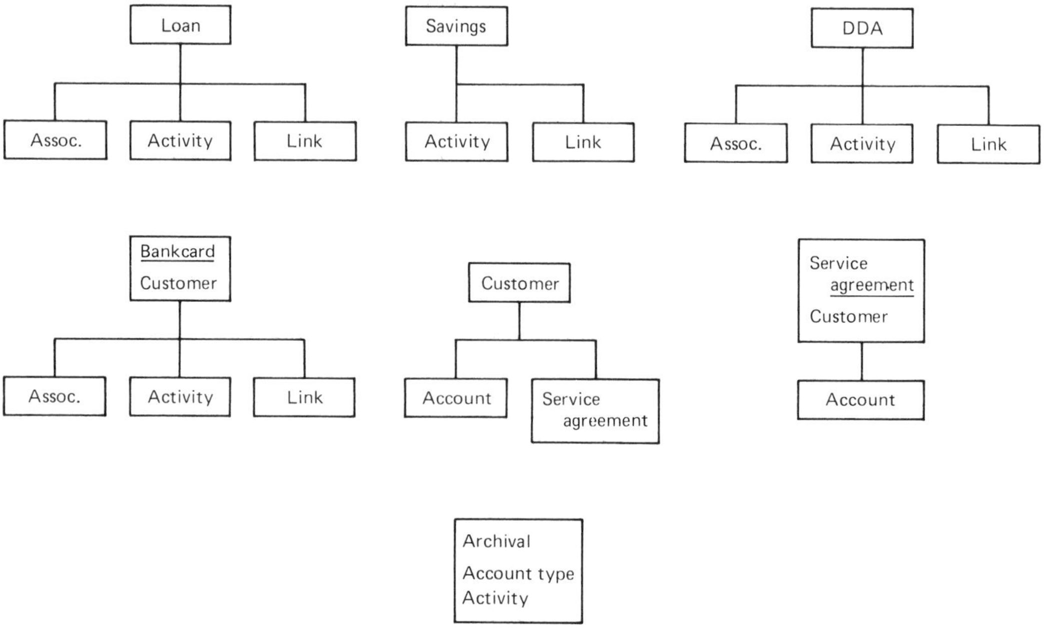

Loan: Assoc. | Activity | Link
Savings: Activity | Link
DDA: Assoc. | Activity | Link
Bankcard Customer: Assoc. | Activity | Link
Customer: Account | Service agreement
Service agreement Customer: Account
Archival Account type Activity

these physical data bases have been designed, decision-support data bases that spring from them can be defined.

The physical data bases satisfy the requirements of the ERD originally specified. Referring back to Fig. 2.18, the customer/activity relationship is satisfied by the four account data bases: loans, savings, bankcard, and DDA. In those data bases each account is linked from account to activity. Note that this and all the other relationships can be satisfied by means of standard hierarchical, relational, or network data base management systems.

The customer/account relationship is satisfied by data in the customer's data base. The account/associate relationship is satisfied recursively by data in the account data base. Note that savings has no associate relationship.

The customer/service agreement relationship is satisfied by data in both the customer and service agreement data bases. Note that a service agreement may have only one customer, but a customer may have multiple service agreements. The service agreement/account relationship is realized by means of data in the service agreement data base.

1. How does the removal off-line of past-dated activities affect performance?
2. Why are smaller data bases quicker to recover?
3. At the stage of physical design, should the designer worry about the size of data bases?
4. What happens when an application designer does not like the plan laid out by the IA team and decides to "do his own thing"?
5. Suggest five different data base design practices that might be implemented to enhance performance and availability for ABC.
6. When should the IA team begin to plan the capacity of the systems that will be built under the architecture that the IA team is building?
7. Can activity data be designed in the absence of other data? Describe what other ABC data there might be and how that data might be optimized for performance and availability.

EXERCISES

1. Translation of the DIS into a physical model begins with strategic considerations.
   (a) What are they? Why?
   (b) What strategic considerations are not made?
   (c) Who should do the translation?
   (d) What if the DIS has not been properly defined—how should resolution be made? How is it usually made?

2. (a) Outline the strategic physical concerns for the following entities (whose DISs have been created in Exercise 1, Chapter 3).

    (1) Account
    (2) Part
    (3) Supplier
    (4) Customer
    (5) Location
    (6) Order

   (b) Define the keys and alternate keys, and give a rough approximation of the physical form of the data.

3. Availability and performance are enhanced by splitting data into small physical units. Why?

4. If I/O was not the constraint it is, how would the design considerations change? For example, suppose that main memory was very plentiful, addressable, and very inexpensive. What would be the constraints on performance and flexibility?

5. The physical model must satisfy more than the primary business-based physical DIS. How can this be done?

6. (a) What are the constraints imposed by storing data on other than direct-access storage, such as on tape, on microfiche, or in main memory?
   (b) In terms of system modeling, what do these constraints imply?

7. (a) What are the criteria for the quality of physical design?
   (b) How does the designer know that a good job has been done? A bad job?
   (c) What is the cost of not doing a good job of physical design?
   (d) Why can physical design characteristics not be retrofitted after the design has been implemented?

8. For one of the DISs in Exercise 2, build a physical model, complete with key definitions, physical characteristics, and so on.

5

Data Views: Bottom-Up Design

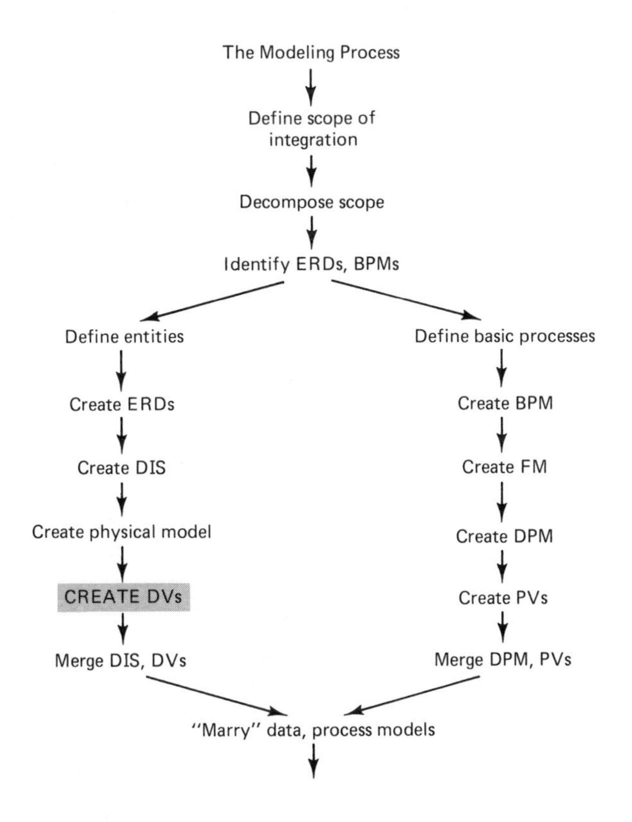

In theory, bottom-up design should produce the same results as top-down design, but it usually does not. There are several valid reasons for this discrepancy. Top-down system design (ERDs, DISs, etc.) addresses:

- The completeness of the system—the fleshing out of the scope of integration
- The importance of the definition of the scope of integration
- High-level entity-relationship diagrams

Bottom-up design, on the other hand, addresses:

- The completeness of a detailed function
- Meshing of several data views to form a comprehensive picture of the system

In theory it is possible to construct a data view for every user, then merge all of the data views together to produce a comprehensive model of the system. But for any but the smallest, simplest systems, such an approach is naive. The enormity and complexity of gathering and amalgamating many data views precludes the project being done properly. Figure 5.1 illustrates symbolically one reason why. Here it is seen that bottom-up views are usually redundant in many respects, but are never quite the same. Also, there are generally many more data views than are effective (i.e., after the first few data views are collected, additional views add very little to the total collective view). Another difficulty with data views is that they represent the world as it is now, not as it *will* be. Data views are generally constructed from a very low level of user—the actual workers. Although this low level of participation is needed to achieve the detail that is desired, a missing ingredient is the future direction of systems. People that are wrapped up in the day-to-day details of today's system usually do not participate in the plans and designs for the future.

Figure 5.1

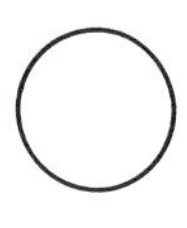

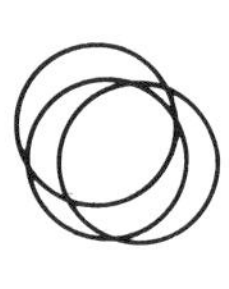

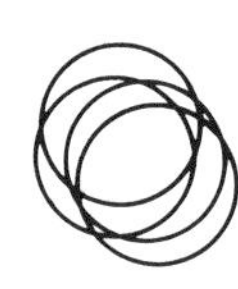

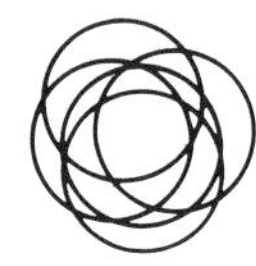

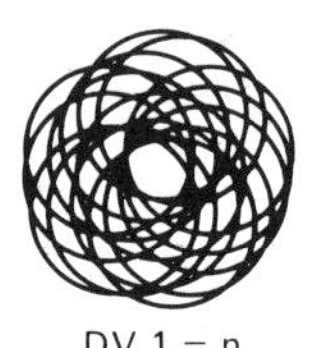

Another difficulty with data views is that they do not lend themselves to as clear a division of different dimensions as is possible with a carefully constructed top-down view of data. Different dimensions tend to become cluttered and confused in a data view representation because of the conglomeration that occurs when data views are amalgamated.

To minimize the overhead, duplication of effort, and red tape associated with a bottom-up construction of *all* of a shop's data views, a better approach is to create strategic data views. The strategy is to create data views in a few, important places, rather than to create all the data views that there are. Some criteria that might be used in determining whether a view is strategic or not are:

- *Representative view:* a view that represents a typical view
- *Criticality:* a view that represents a critical point (in terms of amount of data, timeliness of the data, accuracy of data, focus of data, etc.)
- *Encompassing view:* a view that takes into account much or all of the data being modeled
- *Shift in dimension (operational to decision support, for example):* a view that represents data immediately after it has undergone a dimension shift

The result of a carefully considered set of strategic data views being constructed is an *efficient and effective* definition of the detail of the system. Overlap is minimal and sufficient detail is gathered efficiently.

STEPS IN CREATING A DATA VIEW

Data views can be constructed in many ways. The usual technique is to bring together several end users of the system and ask them how they perceive the system. A typical user-view session takes several hours. The following set of steps represents a consensus:

- Which users does the data represent? What dimension does the data view represent?
- Describe the function being modeled: the input, output, and an outline of the processing that is done.
- Identify the major entities involved in the process.
- Describe the relationship of the entities.
- Define the entities.

(*Note:* The preceding three steps are iterated until they are "clean," that is, until *all* relationships are identified and *all* definitions of the entities are complete, concise, and precise.)

- Define the relevant attributes of each entity.
- Normalize the entities and attributes (assure that attributes are dependent on the key, the whole key, and nothing but the key).
- Stress the data view (i.e., analyze the data from as many perspectives as possible—e.g., is the data view properly constructed from the standpoint of each entity participating in the data view?).
- After user concurrence is obtained, add other relevant characteristics.

The steps used in creating a data view are a microcosm of the steps used to create a top-down design, except that the steps are applied over a much more limited scope. The essential differences between top-down design and data view creation are that:

- Data views are created in a few hours.
- Data views represent the perspective of one user.
- Data views encompass more detail and a much smaller scope.

MERGING DATA VIEWS

Once the data views have been created for as many users as is deemed strategically important, they must be merged. This merging process is represented symbolically in Fig. 5.2. The merging process in essence creates a composite view of the data. Each of

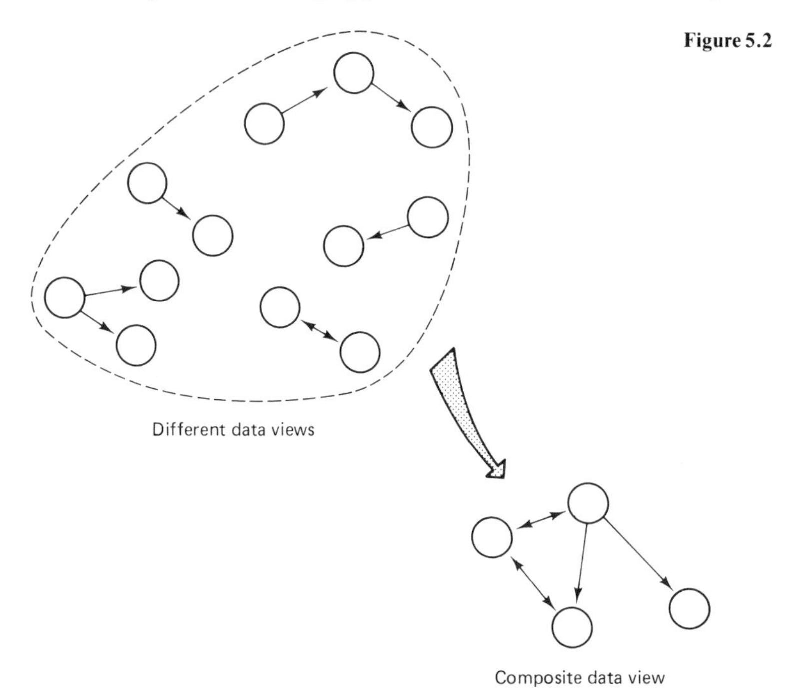

Figure 5.2

the separate data views contributes its entities and relationships to the composite data model.

It is normal for there to be conflicts in the merging of data views. The conflicts usually arise from one of two sources:

- An error
- A shift in dimensions

When an error has been committed, it should be corrected. The more normal cause for conflict in the merging of data views comes from a shift in dimensions. For example, consider the two data views shown in Fig. 5.3. One view is that of the order processor, who must be concerned with the parts that will fulfill an order. The other view is that of an accountant, who must manage the month's orders that have been received in a shipping office. The different dimensions are handled by including in the orders attribute list all those data items that are needed to fulfill both views.

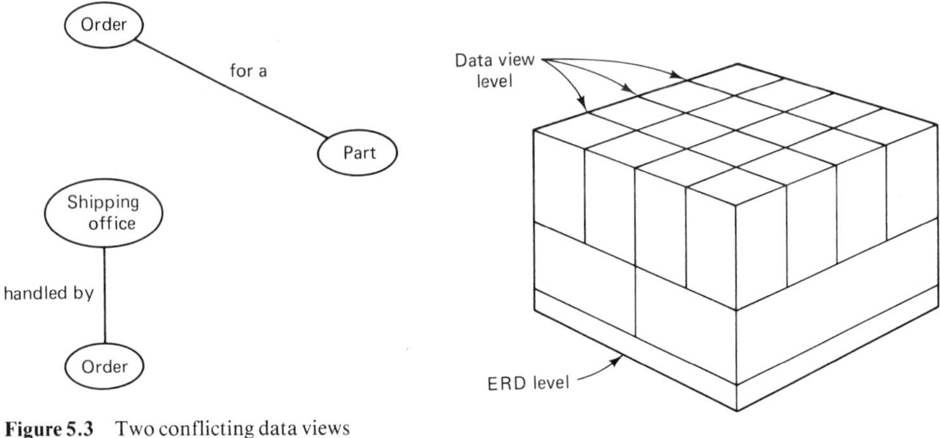

Figure 5.3 Two conflicting data views (from two different dimensions).

Figure 5.4

TOP-DOWN AND BOTTOM-UP MODELS

The differences in top-down and bottom-up modeling are such that *both* techniques are useful in building an effective and stable business-based data model. Top-down modeling is mandatory and bottom-up modeling is well advised. A complete and detailed model can be defined by combining the two processes. There is another perspective of top-down models (as represented by ERDs) and bottom-up models (as represented by DVs) and that perspective is shown in Fig. 5.4, where an ERD is seen as encompassing (or providing a foundation for) *all* of the data model, while DVs are seen as filling in the detail. However, both ERDs and DVs are necessary to complete the full model. The figure also shows that the foundation—the ERD—must be done as the first item of business, while DVs can be filled in as each phase is developed.

CONSOLIDATING DVs: AN EXAMPLE

The DVs shown in Fig. 5.5 represent the views of materials requirements processing, the bill of materials processor, and engineering change processing. They are the results of user-view sessions, as described earlier.

The consolidation of these views is shown in Fig. 5.6. The consolidation is done by merging the different requirements of the data as reflected by the individual DVs. Part number has been replaced by assembly, since each assembly is represented by a

Figure 5.5

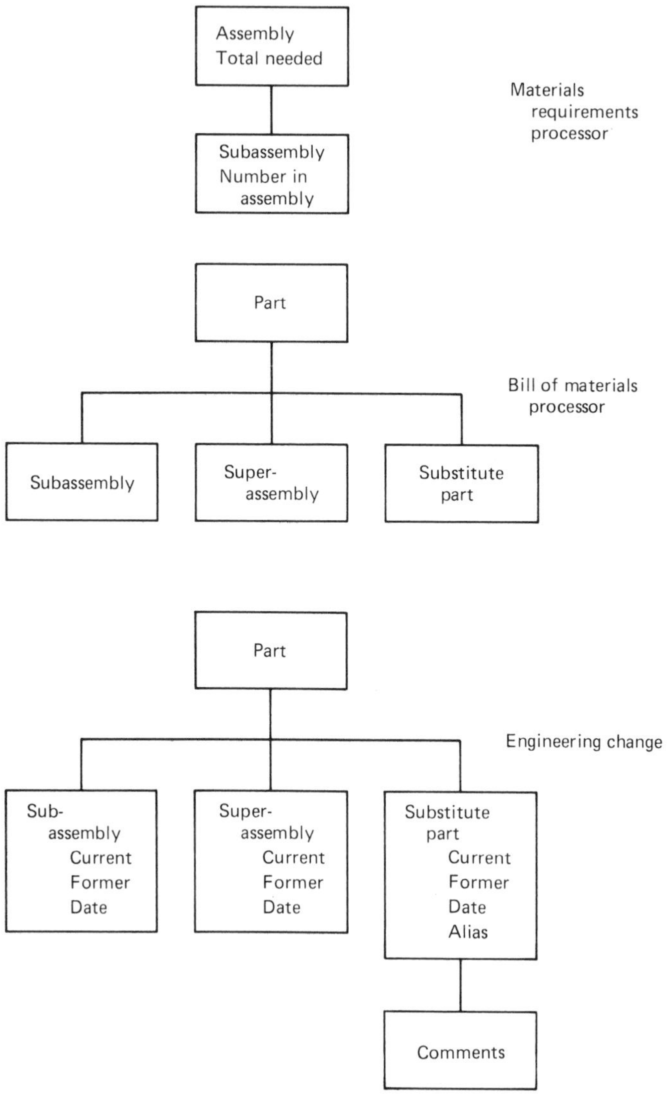

Assembly
Total needed

Subassembly
Number in
assembly

Materials
requirements
processor

Part

Bill of materials
processor

Subassembly

Super-
assembly

Substitute
part

Part

Engineering change

Sub-
assembly
Current
Former
Date

Super-
assembly
Current
Former
Date

Substitute
part
Current
Former
Date
Alias

Comments

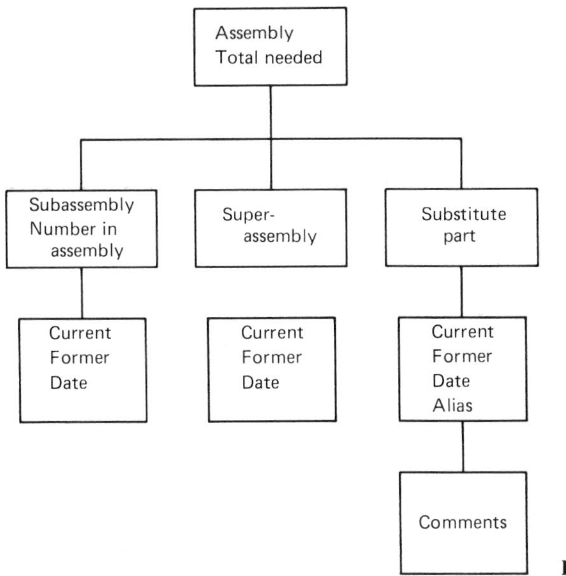

Figure 5.6

unique part number. The current, former, and date data of an engineering change are tacked on to the connector data of the bill of materials processor. In the case where there is no engineering change data, it will not exist dependently. In short, *all* the requirements of the data are satisfied. Note that there has been no essential change in the contents of the data, even though three different data views were modeled. Had there been a radical change in dimensions, data reconciliation would have involved different techniques of consolidation.

CONSOLIDATING DVs: ANOTHER EXAMPLE

Consider the two data views shown in Fig. 5.7. The first view shows data as perceived by a clerk as an account is being opened. The clerk needs to know the name, type of account, and whether anyone else will be associated with the account: wife, husband, son, father, and so on. The second data view is that of a clerk as money is being withdrawn from an account. The essential elements here are account, amount requested, and identification. The first DV is from the administrative side of the system—the auxiliary work that is necessary to set up the system. The second DV represents a part of the normal operation of the system—a withdrawal.

The consolidation of data views is shown in Fig. 5.8, where the associate data has been consolidated into a recursive relationship. The other data is collected and consolidated in the usual fashion. But there is *one major difference* between the DV and the ERD models, and that difference is that the customer is from a different dimension. All the data other than the customer is from the account dimension, whereas customer data is from the customer dimension.

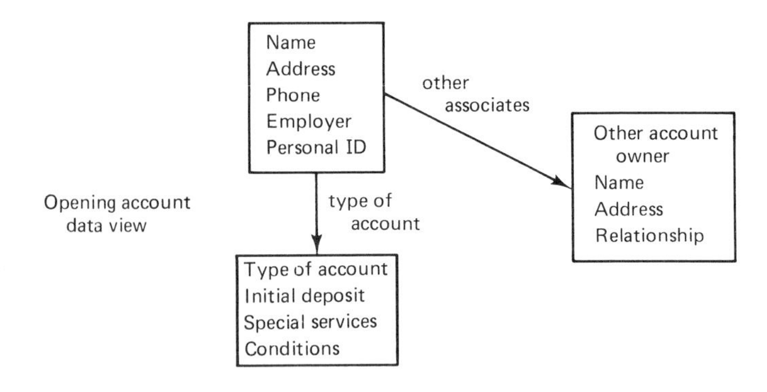

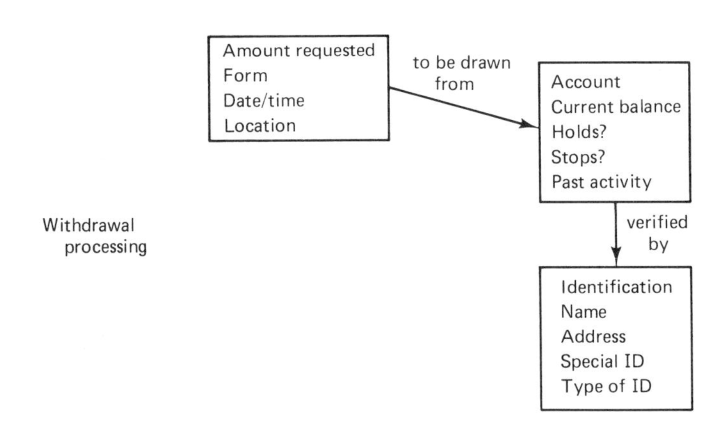

Figure 5.7

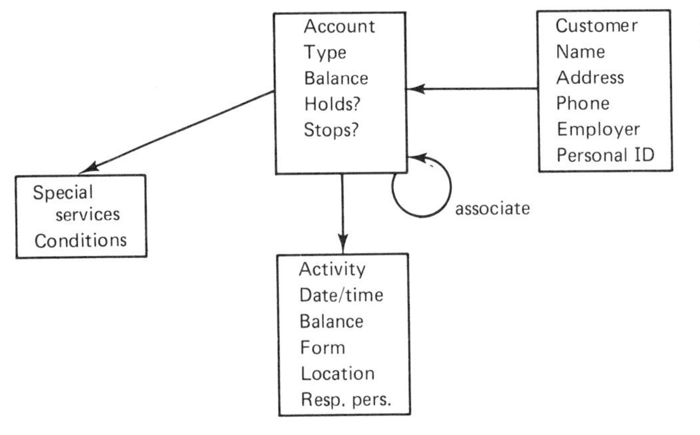

Figure 5.8 Composite of two data views with seemingly contradictory views of the same data.

IN SUMMARY

The process of going from the scope of integration to the physical model is a top-down process. It ensures the completeness of the scope of integration and the compatibility of the ERDs that make up the scope. But a different perspective—a bottom-up approach—is needed to make sure that the data model is complete at the most detailed level.

It does not make sense to model all views of data because there is a high degree of overlap between views. Also, data views traditionally reflect systems as they are today, not as they will be tomorrow. Once several strategic data views are collected, they are consolidated into a composite data view.

PROJECT STUDY

Once the physical data bases are designed (but prior to their implementation) ABC decides to verify that the design is adequate. Using the road map found in the Insert, ABC uses a bottom-up verification of the design by means of the creation of data views. Since there are many data views of ABC's financial systems that apply and most of the views are similar, ABC decides to create views selectively. ABC chooses the view of the teller, the view of vice-president of planning, the branch manager's view, the view of the customer as seen through ATM usage, and the view of the external auditor. (The data views that are collected include much more than activity data, but for the purposes of this project study, only activity data will be included, owing to space limitations.)

In each case the ICO selects a representative sampling of people to participate in the data-view creation sessions. To create the teller view, three tellers (one from a central Denver branch, one from a small remote branch in Glenwood Springs, and one from Wyoming) are called in. A supervising teller is selected, as well as the vice-president of operations.

This diverse group that represents the views of the teller creates its view of data with the assistance of the ICO. From the view created it is found that the existing physical data structures will satisfy most needs. But it is found that interstate activities (e.g., a check domiciled in Colorado but executed in Utah) require a federal transport code when the amount is over $1000. The teller in Wyoming brings this need to the attention of the ICO.

The supervising teller determines that there needs to be a trace facility on a daily basis of the activities handled by a teller. This facility can be realized by the usage of the medium time and date stamp field. In general, the medium stamp is used only by ATMs and automatic transfer devices. Tellers at a window do not use the stamp. But when a teller registers an activity the teller's device can be coupled with an internal sequential count to create a facility for tracing. But this multiple use of a field is contrary to the sound practices of data design, so the ICO decides to create a separate field that will only apply to tellers, rather than to violate the integrity of definition of data.

The vice-president of operations approves the design from the standpoint of satisfaction of functionality. The one question raised by the vice-president is whether all commercial accounts will be treated as being of high volume. Quite a few commercial accounts (doctors, lawyers, etc.) experience a relatively low number of activities. Yet if they need to access their data (e.g., determine if a check has cleared) they must wait in line with long commercial accounts that may require several days to process. The vice-president suggests that commercial accounts automatically be treated as if they were of high volume, but that selected commercial accounts that are of low volume be allowed to be placed on the noncommercial data bases. To accomplish this requires an identification field to be placed with an account's data to separate high-volume and low-volume commercial accounts. This change causes the ICO once again to revise the format of the activity data.

From the creation of a teller's view of data, it has been verified that most functions, but not all, will be satisfied by the physical data base design. The data-view session has added necessary detail to the design.

The other data-view sessions are conducted and similar refinements to the data structures are made. The result of the data-view sessions is a physical design that both satisfies detailed requirements and fulfills the scope of integration.

1. What groups should be represented in a data-view session?
2. How many people should attend a data-view session?
3. What happens if a key person is unable to attend a data-view session?
4. What happens if a bottom-up developed data view conflicts drastically with the physical model? With another data view? How can the conflict be resolved?
5. Does the physical model necessarily imply the use of a particular DBMS? Why or why not?
6. What levels of personnel should attend a data-view session?

EXERCISES

1. (a) What is a data view?
   (b) How does it differ from a dimension?
   (c) Who creates data views? Who participates in the creation of data views?
   (d) What advantages does data-view creation have over top-down design?
   (e) What are the differences in the design produced only by consolidating data views and the design produced by top-down design?
2. Why is a data view more "fleshed out" than a corresponding DIS?

3. Create data views for the following people:
   (a) A bank teller working on a savings account
   (b) An accountant in the manufacturing environment
   (c) An insurance agent selling life insurance
   (d) A football player doing calisthentics and warm-ups
   (e) A TV announcer preparing news copy
   (f) A policeman patrolling a dangerous neighborhood
4. Why is it inefficient to collect all data views and mesh them together? What are the risks that the resulting consolidated view will not be accurate?
5. (a) Why are different modes of operations and dimensions difficult to recognize at the data-view level?
   (b) Why are they easier to recognize from a top-down perspective?
6. Compare and contrast the steps in building DVs and DISs. Where are they identical? Similar? Different?

6

Merging Top-Down/Bottom-Up Data Models

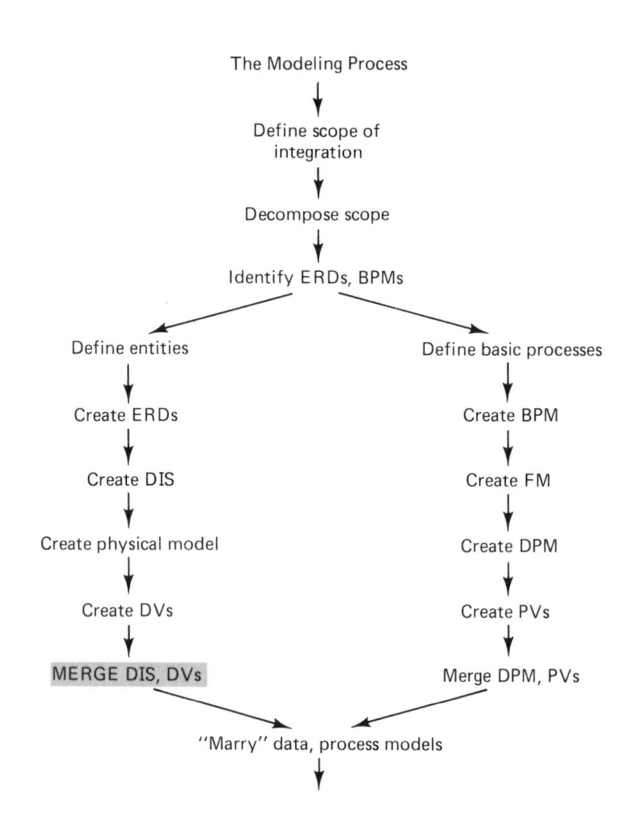

The normal progression in the development of a business-based data model is to define the scope of integration, define the ERD that is appropriate to the first phase of development (usually the primary business-based ERD), and then refine the ERD into DIS and data views. At this point one major view of the system (the primary business-based view) is fairly well defined but other major aspects of the system that lie within the scope of integration are at best sketchily defined. System definition has proceeded along a singular functional train of development, as shown symbolically in Fig. 6.1.

As an example, suppose that the scope of integration were the retail systems of a bank. The ERD might have included accounts, customers, services, and activities. The entities selected for detailed enhancement are categorized into loans, savings, checking, and so on. The DISs are built and verified by the creation of DVs. Although the primary business-based ERD is now fleshed out, there are still major areas of banking retail systems that are untouched. The establishment of accounts and service relationships, the impact of the organization structure on the systems, and the auditing of accounts, for example, are not represented by the primary business-based ERD, and yet these areas are within the scope of integration. To be complete, an ERD must be created for each of these areas and must be integrated with other existing ERDs into the same scope of integration.

After the first model has been built, the next step usually is to "fill in" the rest of the model (i.e., complete the scope of integration). At each level of modeling (ERD, DIS, DV) the scope of integration is expanded. This expansion accomplishes

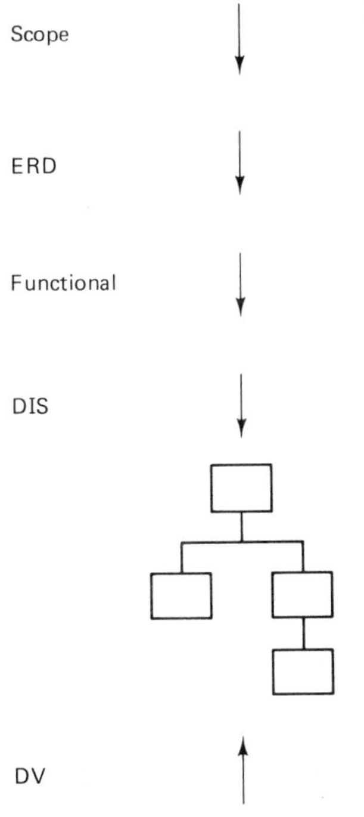

Figure 6.1 First path in building the data model.

two things: The scope of integration being modeled becomes more complete, and the different interrelationships of different components of the model at different levels become defined. The completion process is shown symbolically in Figs. 6.2 and 6.3. Figure 6.2 shows that once the scope has been determined, ERDs can be defined. From the ERDs, DISs can be defined. The physical model can be created after the DISs are defined and consolidated, and DVs are created to verify the completeness of detail of the physical model. Figure 6.3 shows the results of the modeling effort. There is a well-defined scope of integration, one or more ERDs, many DISs, physical models of data, and composite data views.

Because of personnel limitations and the timing of phase development, the data model may take a long time to complete. Of course, the portions of the data model central to the immediate success of the business are completed first and those portions of the model tangential to success are completed last. In the case where the tangential portion of the model is vital (in some respect) to the primary business-based data model, the tangential portion is completed to the extent that future development will not have a negative effect.

For example, suppose that the audit function is vitally important to the business (but unimportant insofar as making a profit is concerned). Then the audit function

Figure 6.2 Filling in the data model.

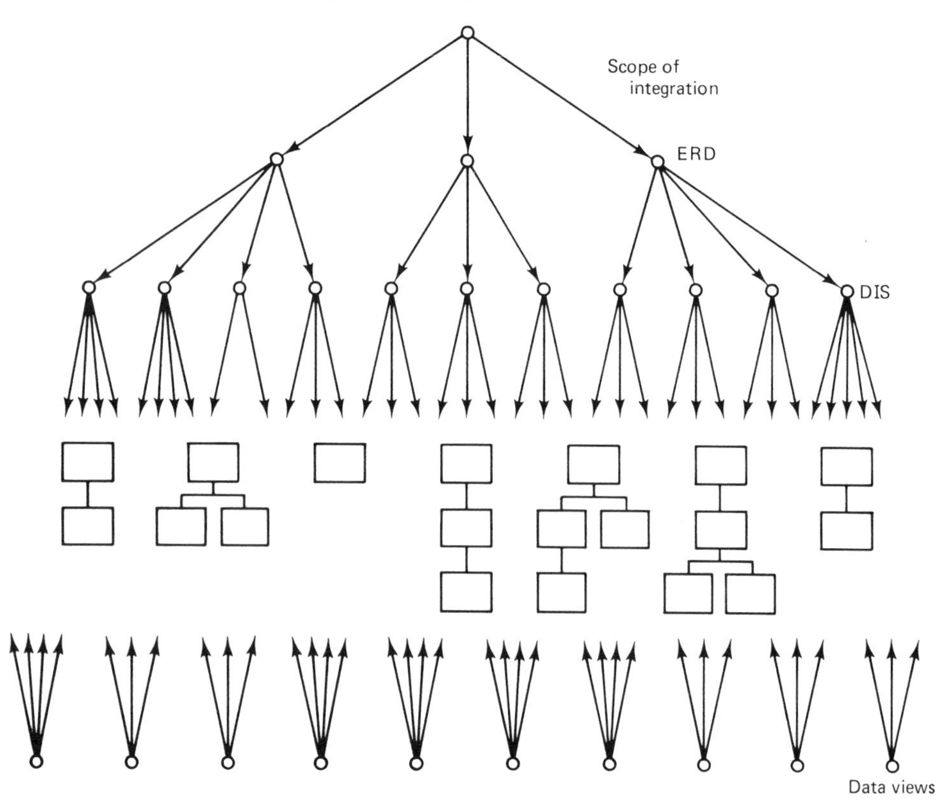

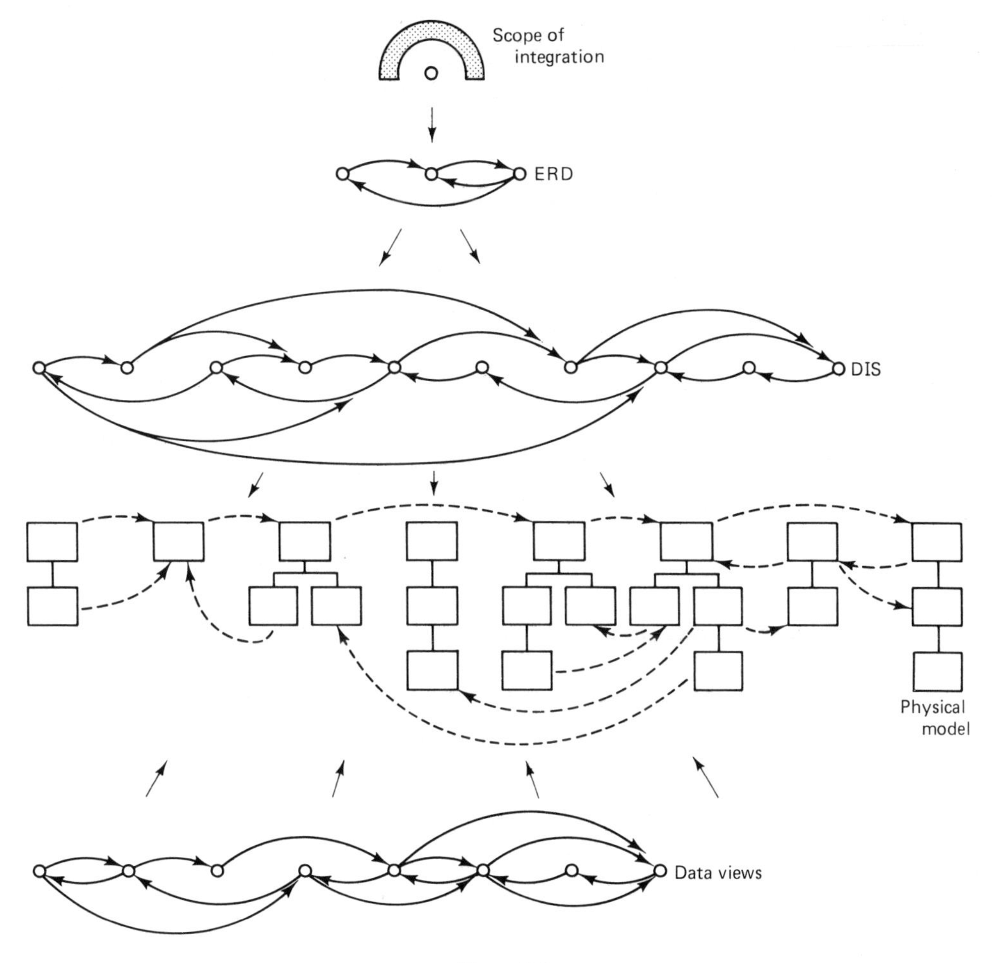

Figure 6.3

must be modeled to the extent that its interface throughout the business is defined (i.e., audit trails, audit tools, and so on, are identified). When this is done, the primary business-based model may be developed actively and the audit portion of the system may be fully developed at a later time.

Of particular interest in the completion of the data model are the following activities:

- Building the global ERD
- Merging dimensions
- Building entity dimension maps
- Merging data views and data item sets
- Establishing model connections at each stage
- Resolving conflicts

GLOBAL MAPS

Two kinds of global maps can be built, depending on the size of the organization being modeled. One map is the global map of the scopes of integration, shown in Fig. 1.15. The global scope of identification map identifies all the various scopes of integration for an organization and shows how they are related. This map is drawn at a very high level. For example, a bank may have a retail banking scope, a landlord scope, a software licensing scope, and so on.

The second type of global map is the global ERD map. The global ERD map describes all the dimensions within a given scope of integration. The primary purpose of the global ERD map is to identify interfaces between different dimensions within the same scope of integration. The map also shows what dimensions must be included in the various entity dimension maps that will be built for each entity.

For example, suppose that the scope of integration comprises the financial systems of a bank. It must be possible to identify a bank's financial customer whether the person is in the loans system, the savings system, the bankcard system, or wherever. The customer would appear in each of the dimensions, and each dimension would ultimately be related to another dimension. The global ERD map shows the relationships of the various dimensions with the scope. Because the global ERD map is inherently complex, as many details as possible should be stripped from the map. An example of a global ERD map for the operational systems of a manufacturer is shown in Fig. 6.4. Note that *only* the entity name and relationship are identified. The global

Figure 6.4 Simple global ERD for the operational systems of a manufacturer.

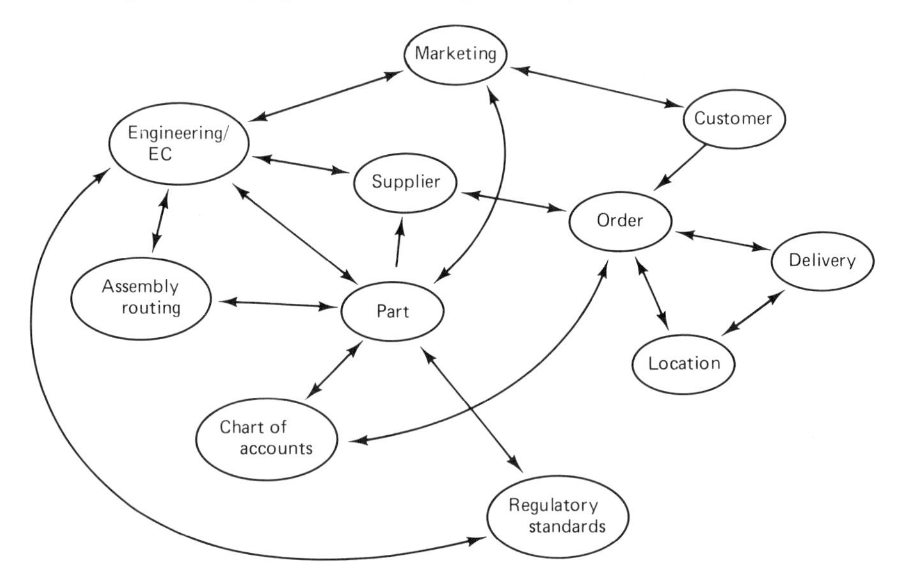

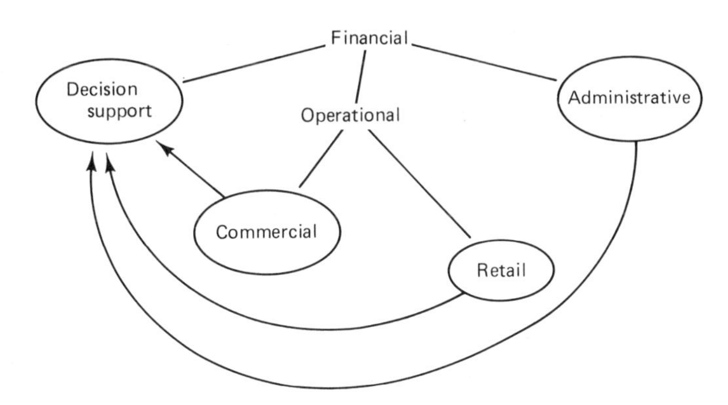

Figure 6.5 Global scope of integration map for a bank's systems.

ERD map represents the summation of *all* dimensions within the scope, and as such represents a three-dimensional diagram of data (which accounts in part for its complexity—for further discussion of the three-dimensionality of integrated data, refer to the discussion of the chandelier perspective of systems in Chapter 6 of *Integrating Data Processing Systems: In Theory and in Practice*).

The global scope of integration map usually represents different modes of operation and as such is used only to specify interfaces. For example, Fig. 6.5 shows a connection between different scopes: decision support, commercial, retail, and administrative. It will be supported by its own distinct, yet related system. It will be necessary to send data to decision-support systems from other scopes. The data sent to the decision support scope of integration needs to be understandable in terms of keys, formats, and so on. The global scope of integration is used to point out these interfaces.

MERGING DIMENSIONS

As a shop begins to fill in the data model of its scope of integration, more than one dimension is developed. Usually, one or more entities participate in multiple dimensions. There is no conflict here because an entity (at the ERD level, DIS level, or physical data model level) may contain connectors and attributes that apply to more than one dimension. Consider the DIS shown in Fig. 6.6. In this case, part (or product) is shown in its form from four different dimensions: that of the inventory clerk, order clerk, accountant, and engineer. The differences in perspectives (which serve to reinforce the concept that data is viewed in at least three dimensions) is resolved by an amalgamation of data to satisfy *all* viewpoints.

One convenient way to show the total dimensional requirements for an entity is through an entity dimension map.

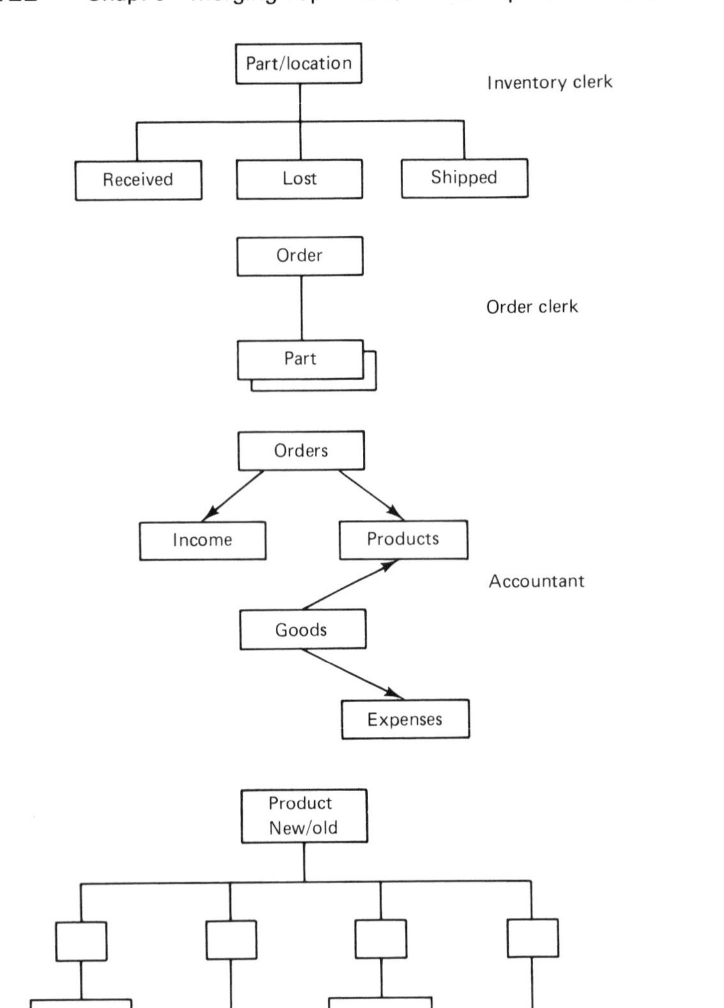

Figure 6.6

ENTITY DIMENSION MAP

An entity dimension map is formed by the collection of the different dimensions in which an entity participates. This map is usually created and used informally. For the sake of simplicity it makes sense *only* to relate a single entity to *all* the dimensions that it touches. In other words, an entity dimension map is not a view of data from any one perspective, but an accumulation of all the different ways in which a single entity can be viewed within a single scope of integration. This amalgamation is necessary to determine the total business requirements of the data. For the purposes of the entity dimension map, no other data should be included in the map other than the dimensions that relate directly to the entity being modeled. Another point of interest is that even though an entity dimension map is drawn at the entity level, the repercussions on the model are found at *all* levels. Ultimately, the physical model must take into account all dimensions in which an entity participates. As an example of an entity dimension map, refer to Fig. 6.7.

When all dimensions are not included that ought to be included in the entity dimension map, the resulting data model will be incomplete. Later, after the system is already constructed, either the missing data requirements will have to be added after the fact or certain parts of the data model will have to be physically recreated. In the case of data element additions, the addition would be unnecessary if the model had been created properly from the outset. In the case where new data models (and new systems) are created separately from the original data model (and original system) redundancy of data and processes creeps into the design and defeats one of the major purposes of data modeling.

For example, in looking at Fig. 6.7 a data model is created and systems are built based on the model. But suppose that a dimension—marketing—is discovered after the system has been built. Marketing requirements for the data have not been considered at all by the model in Fig. 6.7. There are two courses of action: to go into the existing model and existing systems (which may or may not be easy to do) and make changes, or to create a separate product model for marketing (which for all practical purposes defeats the purpose of modeling in the first place!). If two separate models are created—a marketing model and a manufacturing model—there certainly must be some degree (probably large) of redundancy of data and processing.

Figure 6.7 Entity dimension map.

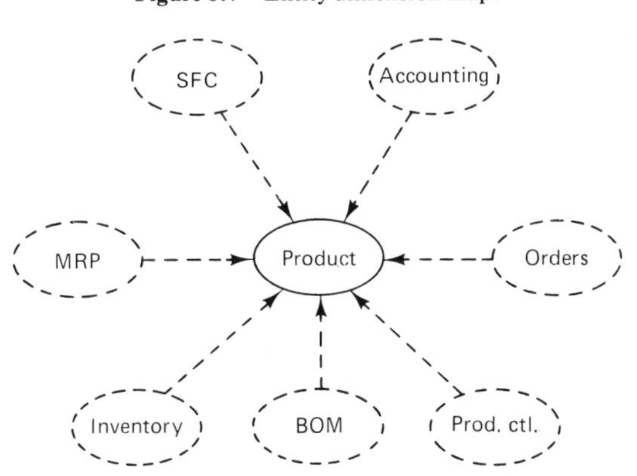

MERGING DATA VIEWS AND DATA ITEM SETS (TOP-DOWN/BOTTOM-UP RECONCILIATION)

Once the data item sets are created, the usual practice is to strategically select data views that are necessary to complete the details of the data model. Once the data views are constructed, they are ready to be merged with the corresponding DIS or consolidated DIS.

The most effective practice is to consider the merger of DIS and DVs from the viewpoint of the entity dimension map. Such a map is illustrated in Fig. 6.8, where an entity dimension map has been used to identify what DIS should be used to relate DVs to the product. The designer began with a product. Then from the product the entity dimension map showed what dimensions should be accounted for (i.e., BOM, MRP, SFC, accounting, order, and inventory). Next, the designer retrieved the DIS for each of the related dimensions and created the model shown on Fig. 6.8. This figure shows the modeler what relationships the model should satisfy at a very specific level. Note that *only* the relationships from other dimensions to the product are considered (i.e., the other relationships between orders, accounting, BOM, etc., are not considered in the building of the model at this point).

To merge data from the DVs with the DIS, each DV is compared with the corresponding dimension in the entity dimension map shown in Fig. 6.9. If a data element appears in the DV but is not in the DIS, the element is added to the model. This is another way in which detailed data is added to the model. Note that data is included in the model even when it does not specifically apply to the product. For example, data is added to the orders model based on the details discovered by creating a DV for the product. When it comes time for an entity dimension map to be drawn for orders, the detailed data will already be included. The merging process is depicted in Fig. 6.9.

After the model is completed for the product, another entity dimension map is created for the next entity to be modeled until all entities have been modeled.

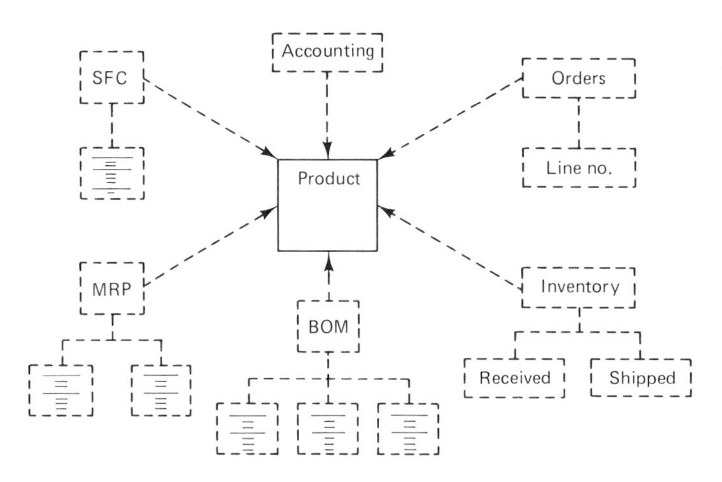

Figure 6.8 DIS model expansion of the entity dimension map.

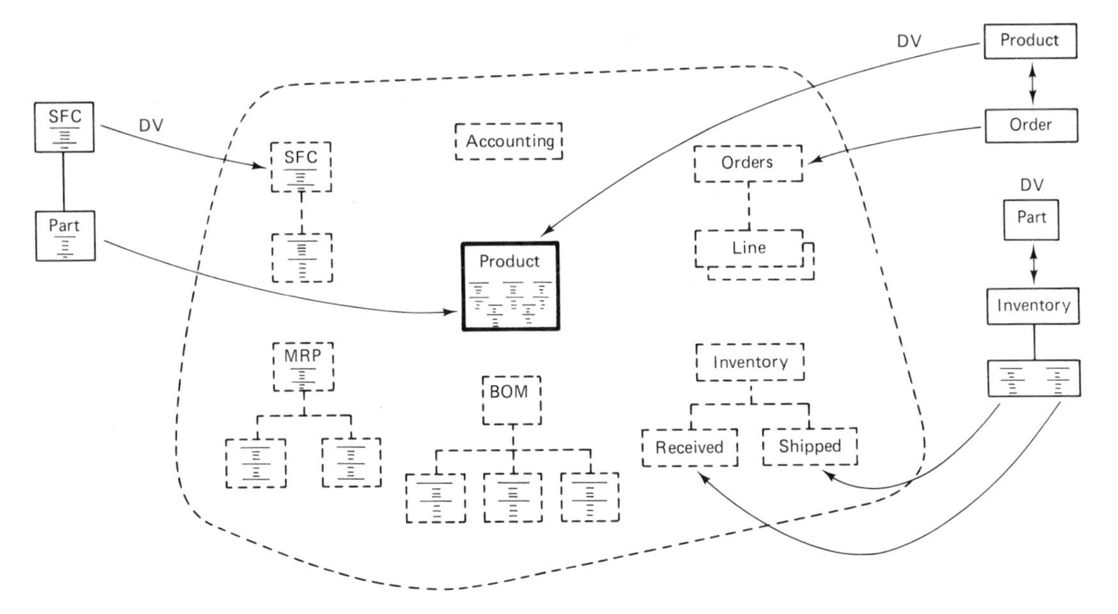

Figure 6.9 Entity dimension map at the DIS level. All data views are collected and are matched against the corresponding DIS model to make sure that the data view can be satisfied. If the data view cannot be satisfied, data is added to the DIS model.

ESTABLISHING CONNECTIONS IN THE MODEL

Throughout the building of the top-down model, connectors have been established to show that data externally existed and was related, but existed outside the dimension of an entity (or DIS, or at whatever level was being considered). As part of the process of DV and DIS reconciliation, the ability to relate data from different dimensions should be verified. At the point of data modeling, the issue is: *Is* there a connection between different parts of the data model? At the point of physical modeling, the issue will become: How efficient is it to use the connection? Once data views are reconciled against the DIS model, the physical model is ready to be built.

RESOLVING CONFLICTS

For the most part, the process of merging DVs and the DIS model is straightforward. But on occasion there arise conflicts that are seemingly without resolution. Conflicts can be classified to explain why they arise and consequently what can be done about them.

Simple Conflicts

Simple conflicts show up as incomplete data or relationship requirements—a DIS is missing some attributes, for example. The solution is to merge requirements and enlarge the attribute list of the DIS and the physical model. This conflict resolution occurs for data within the same dimension. This type of conflict and resolution represents about 90% of the conflicts faced by the system architect (if the system is methodically and carefully constructed).

Complex Conflicts

Complex conflicts appear as a need to organize data completely differently. They occur because a dimension within the scope of integration has not been considered. The solution is to include and integrate the new dimension with the other dimensions within the scope of integration. This conflict is entirely resolvable within the scope of integration. If the scope has been carefully defined at the start, this type of conflict arises infrequently. If the scope has not been defined (either carefully or at all), this type of conflict arises frequently.

Very Complex Conflicts

Very complex conflicts appear as a need to have a completely different organization, content, and usage of data. This type of conflict is outside the scope of integration and is not easily resolved. The only real long-term solution is to redefine the scope, but this is usually not feasible or desirable, for many reasons (timing, resources, undoing existing systems, etc.) The usual resolution is to build another scope of integration. The magnitude of this level of resolution is very, very large and should be avoided at all costs.

The most profound conflicts often occur because major shifts in mode of operation have not been included in the scope of integration at the outset. Typical of these modes are:

- Operational versus archival
- Operational versus administrative
- Operational versus decision-support systems

Another cause of major conflicts are shifts in business. For example, if a manufacturer attempts to combine the business of car rentals, it is a good bet that the manufacturing model of data will not fit the car rental business.

IN SUMMARY

The first pass through building the data model normally centers around going through a single pass of all levels of modeling activity for the primary business-based model. Once that model is built, the scope of integration must be filled in. To do this, global ERD maps, entity dimension maps, and merged dimensions must be constructed. After the consolidated DISs are built, DVs are mapped against them to determine data content and relationships. The DISs ensure that the scope of integration is complete; DVs ensure that the detail at the point of operation is complete. The two are necessary to ensure that the resulting physical model of data will be adequate.

EXERCISES

1. (a) What are the steps in the "normal progression" of system modeling?
   (b) What happens if a step is omitted? *Can* a step be omitted?
   (c) Does the primary business-based ERD and DIS have to be developed first?
   (d) What other options are there?
2. (a) What is meant by "filling in" the model?
   (b) When should it be filled in?
   (c) Does it have to be filled in? What are the consequences of not filling it in?
3. Create an entity dimension map for the following:
   (a) A part
   (b) An account
   (c) An insurance policy
4. (a) What happens if one or more dimensions are omitted from the entity dimension map?
   (b) What happens if more than dimensions are included in the map (i.e., entities other than those that relate to the entity being mapped) are included?
5. (a) How does the chandelier perspective (as described in *Integrating Data Processing Systems: In Theory and in Practice*) relate to the entity dimension map?
   (b) In what ways is an entity dimension map more descriptive of data than the chandelier perspective? Less descriptive?
6. (a) Why is a global ERD map generally *less* useful than a local ERD map?
   (b) Is there a scope of integration that covers the data in a global ERD?
   (c) Can the same data be represented in more than one ERD that exists in the global ERD map?
7. (a) What difficulties does one encounter in merging DVs and DISs?
   (b) How can the conflicts be resolved?

7

Process Modeling: The Basic Process Model

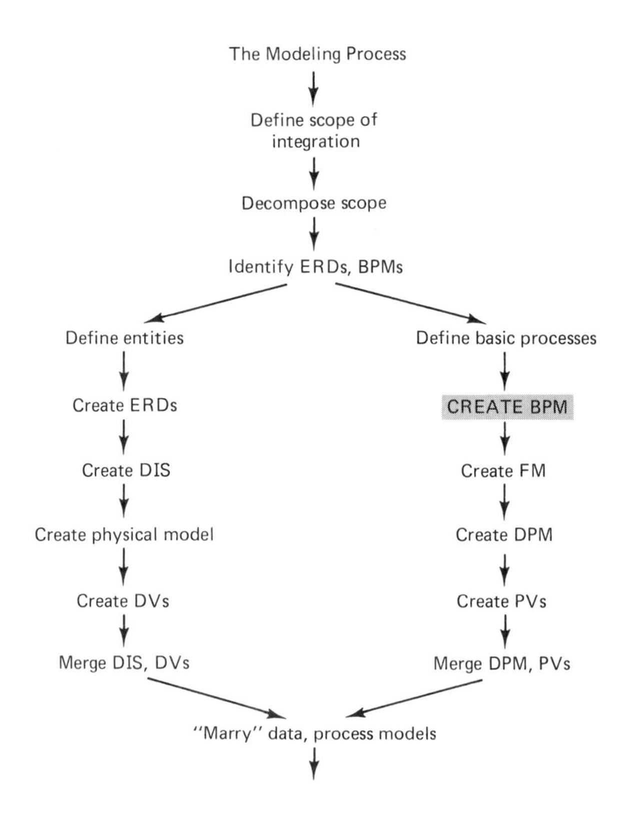

If a strong case can be made for carefully modeling the data of a corporation, an equally strong case can be made for modeling its processes. In fact, *both* types of models are necessary to form a complete picture of the corporation's systems.

Traditionally, the techniques of data modeling have been applied to the on-line environment, where data is shared by more than one user. It is easy to see the need for a common model where data must satisfy more than a single set of requirements. By the same token, process modeling has long been recognized in the batch environment as a tool to describe fully the processing of a user. Structured techniques of analysis and design suffice to describe the batch user's needs.

Historically, shops have built one or more forms of process models, usually based on structured techniques. When building the process model, if such previously built analysis has already been done, and if it is accurate, there is no point in completely rebuilding it in the format about to be described. Instead the existing analysis should be analyzed and revised only where necessary, to achieve the same results as the modeling effort produces.

But in a large, integrated environment, where there are batch and on-line processes, and where there is a genuine concern on management's part for the development and maintenance budget, *neither* data modeling nor structured design techniques, in and of themselves, suffice. To come to grip with the *total* set of issues, the business-based system developer must adequately address *both* data and process modeling, and in equal proportion. Without a process model that accounts for a system's process from the broadest perspective, it is very difficult to construct systems without a high degree of redundancy. An unmodeled system tends to be built piecemeal and from a local perspective, where like processes are constructed in many different places. Such a piecemeal construction effort is shown in Fig. 7.1. In this figure, five different systems are built over a period of time. Functions A and B exist as an integral part of each system but are built separately each time the system is constructed. There are other differences in the systems and there are differences in the size of functions A and B, but the essential process does not change.

As an example, system A might be the bank's savings deposit system, system B the money-market system, system C the payroll deposit system, system D the DDA (checking account) system, and system E the IRA (Individual Retirement Account) system. As shown, each system has been built separately. Function A is the tax-reporting function, where the bank reports interest paid to the government. Function B is the account management function, where an account is credited or debited, and so on. As shown, functions A and B are built each time a new system is built, even though the function is identical. In addition to being extremely wasteful of development time, maintenance is unduly difficult, as is the synchronization of data. The cost of development together with the cost of maintenance is at least five times as large as it would have been had functions A and B been built once. But to have built A and B once requires an understanding of *all* the requirements for which A and B will be required and a commitment to build and use A and B once, at the beginning of the project.

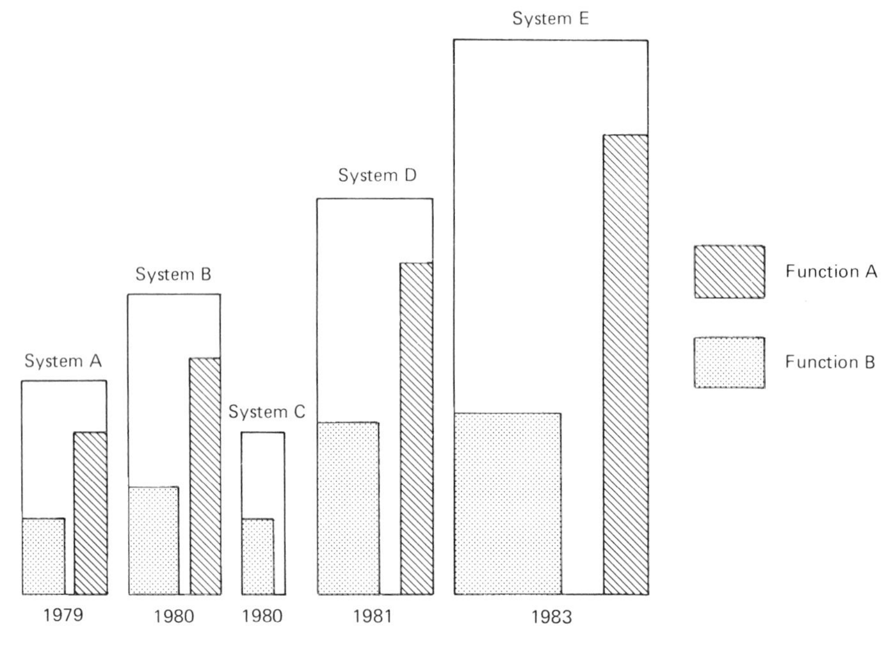

Figure 7.1

The ideal case is depicted in Fig. 7.2, where, functions A and B (the tax-reporting function and the account management function) are defined at the outset to service the entire set of needs of all the systems within the scope of integration, not on a system-by-system basis. In the first system that is built (in 1979—the savings deposit system) are functions A and B. In 1980 two systems (money market and payroll deposit) are built that use A and B. They are built using existing code. In 1981, and in 1982, another system is built. In all, A and B are built *once* and maintained *once*.

But to build A and B so that they can be constructed once requires two essential things:

- A commitment to integrate
- An understanding of how to build integrated systems

Figure 7.2

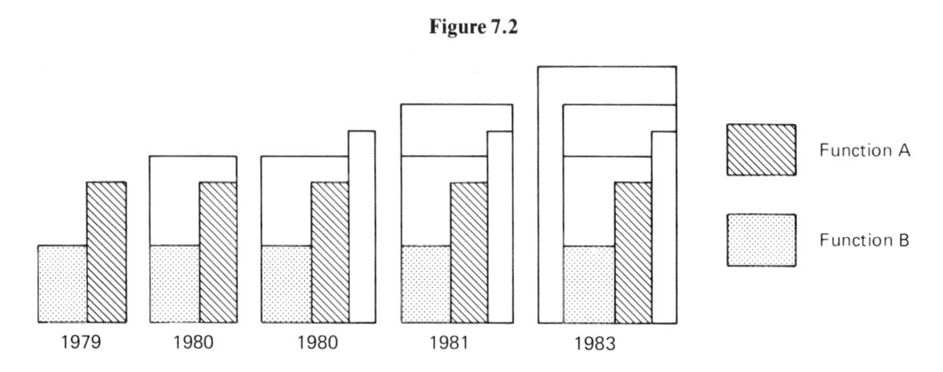

It is clear that integration, done properly, greatly reduces the redundancy of processing and data in a system. The first step toward achieving the integration of processes is to understand each process as it relates to the scope of integration. The most effective way to understand the processes of the scope of integration is to decompose and abstract them, much as data must be abstracted.

As an example of the usefulness of abstraction in analyzing processes, consider an example of the processing to be done for the issuance of a life insurance policy and a homeowner's insurance policy, shown in Fig. 7.3. In automating the life and homeowner's policy issuance process, the designer has at least two choices: to build two distinct systems or to build one system with distinct parts. From a system development/maintenance point of view, building one system with distinct parts is much more efficient, but requires

- More planning
- Timing
- Cooperation between the users

If there were only two systems—life and homeowner's—to be built, the advantage of combining processes is not that much, but consider other lines of insurance, such as

- Fire insurance
- Medical insurance
- Professional liability
- Auto insurance

Is it necessary to build a separate information system for each type of insurance? A billing system? A claim processing system? Of course it is not necessary.

Figure 7.3

Life insurance processing	Homeowner's insurance processing
Contract prospect	Contract prospect
Determine basic life information (name, address, age, sex, etc.)	Determine basic homeowner information (name, address, age, etc.)
Policy type? liability? term? whole? other?	Specifics of policy (rental, structural, etc.)
Determine medical history	Review property – jewelry, art, cameras, electronics, any goods over a set amount
Medical exam required?	Appraisal necessary?
Set rates – factors: age, sex, marital status, occupation, medical history, other	Set rates – factors: total amount of policy, neighborhood, type of dwelling, size of dwelling, occupants, contents, etc.
Set policy limits	Set policy limits
Determine premium	Determine premium, payment schedule
Determine beneficiary	Servicing agent selection
Determine servicing agent	Deliver billing information
Deliver billing information	Record data
Record data	

ABSTRACTING PROCESSES

Consider the issuance of life and homeowner's policies. The first step to understanding abstraction is to divide the processes into two categories: unique processes and common processes.

Unique	*Common*
Determine medical history	Contact prospect
Set rates	Basic information gathering
Set policy limits	Policy-type determination
Beneficiary	Premium, payment schedule
Property review	Servicing agent
	Billing information
	Record data

Common processes should be built once, and built to accommodate unique processing requirements. However, the commonality of processes must be identified and addressed prior to the building of a system, because once a process is built, it serves only the set of requirements for which it is built. It is only happenstance when a process fits requirements for which it was not originally designed.

How can the commonality of processes be determined? There are basically two approaches:

- *Top down:* from the broad perspective of *all* processing (in this case, from the view of each type of insurance)
- *Bottom up:* from the amalgamation of each of the types of processes (in this case, by comparing fire insurance with auto insurance with homeowner's insurance, and so on)

It is not an accident that these two approaches to process modeling have a direct correspondence to the similar approaches of data modeling. Indeed, each approach shares the same rationale and objectives.

- *Top down:* completeness of scope, relationship of major parts, uniform definition
- *Bottom up:* details of a given process

As with data modeling, it is mandatory that the top down approach to modeling data be done, while it is highly advisable that the bottom-up approach be taken and merged with the results of the top-down model.

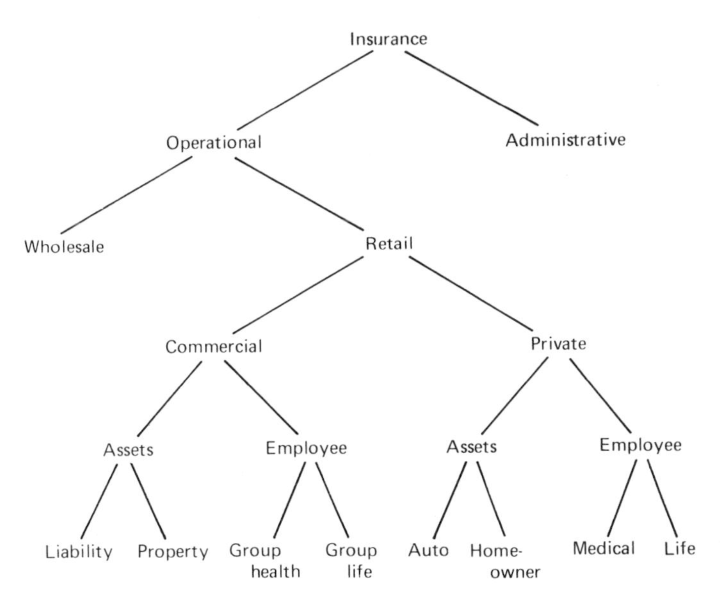

Figure 7.4 Decomposition of the operational retail system of an insurance company.

SYSTEM DECOMPOSITION

As with data modeling, the basic process model (which is being built top-down) begins with a decomposition of the scope of integration. Such a decomposition of insurance processing is shown in Fig. 7.4. The decomposition that is done is the same as that discussed earlier in the chapters on data modeling. The result of the decomposition is the identification of the basic business processes and entities and an identification of the basic cycle of processing of the business.

The decomposition shown breaks up the operational retail systems of an insurance company (which for the purposes of this chapter is taken to be the scope of integration). Even though the organization and types of insurance serviced may be organized as shown by the decomposition, the basic business of an insurance company centers around several common processes that are found in all types of insurance. At each level—commercial liability, group life, personal medical—the business of the insurance company centers around processing policies, claims, premiums, and so on. From the processing common to *all* types of insurance a basic process model (BPM) is constructed.

BASIC PROCESS MODEL

While the technique of process modeling is important, it is easy to become so entangled in technique that one loses sight of why process modeling is occurring. Indeed, *any* technique (structured, etc.) that accomplishes the objectives is acceptable. The objectives of process modeling, like those of data modeling, are:

- Identification of *all* processes within the scope of integration
- Identification of commonality of processes (so that redundancy, at a high level, can be identified and addressed once)
- Identification of the structure of processes (how different processes relate)
- Identification of the details of processing (what processing occurs)

The basic processes for each type of insurance are: buying or renewing a policy, issuing and crediting premiums, handling claims, and reviewing rates, prior to the renewal of a policy. Figure 7.5 illustrates the basic process model. Note that the BPM is as fundamental to the process model as the ERD was to the data model. The diagram describes a basic cycle of processing, where a policy is either bought or renewed, after which a premium is calculated and billed. Of course, the flow of activity in the BPM is based on the actual flow of activities as experienced by the user. If a claim(s) is made against the policy, it is handled (either paid or rejected). In any case, the claim information is stored and is used at rate review time. Periodically (annually, semiannually, etc.), a policy comes up for renewal, at which point the rates are reviewed and adjusted. The policy cycle then begins anew. The cycle holds true for all lines of insurance. (Note that when the cycle is completed, there is a return to the beginning of the cycle. This is true for cycles defined at the highest level. For cycles at a lower level, this is usually not the case.) Of course, the rates set, rate reviews, claims handling, and so on, will differ for the individual types of policies, but much of the basic activity (premium billing, claims initiation, policy renewal, etc.) will not differ.

The level of detail of the BPM is scant. What belongs in the BPM and what does not belong depends entirely on the business of the corporation. The basic process is identified and how it relates to other processes is identified, but other than that, the BPM contains little information. The level of detail of the BPM corresponds to the level of detail of the ERD. In later extensions of the BPM the differences and specifics of the model will surface.

At the highest level of process cycles, many different types of cycles are represented at the highest level of abstraction. For example, the basic cycle identifies the basic process of buying or renewing a policy. This represents the activities of resetting limits, establishing beneficiaries, determining basic coverages, determining deductible amounts, establishing liability, and so on. All of these activities are types of buying or renewal of a policy, and these activities apply across different kinds of

Figure 7.5 The BPM for insurance

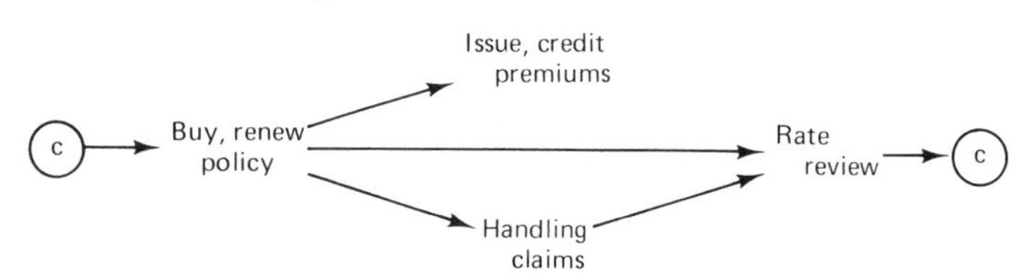

policy. So the analytical tool of "is a type of" applies for the process model as well as the data model.

As a pragmatic side note—the work that is required for analysis of data runs parallel with the same sort of work for processes. There is no reason why the analyses cannot be done simultaneously. Doing so avoids having to reconvene the same people and discuss what is essentially the same topic but with a different slant.

One aspect of the BPM is that it represents activities which are fundamentally different. Within the scope of integration the activities cannot be considered to be of the same type. If any two activities in the BPM can be the same type of activity, they should be combined (as in the case of the entities within an ERD). For example, it is possible to call "buying a policy" one activity and "renewing a policy" another. Although there are differences, the activities are essentially types of "policy issuance." Therefore, in the BPM, policy issuance is appropriate, whereas buying or renewing a policy is not.

After the BPM and the basic business cycle are established, the gross definition of the activities of the primary business-based process are created. Much like the primary business-based ERD, the primary business-based BPM does not include activities that are not directly related to the primary business of the company, such as auditing and policy accrual activities. However, the primary business cycle describes all activities that are needed to make it complete. No major activity is omitted. Detailed activities that are types of major activities will be described as part of the process model, but at a lower level.

Of course, once the primary business-based process model is described, other dimensions that are to be modeled can be described to begin to fill the scope of integration (in much the same way that multiple ERDs filled in the data model). Once the process model has been established, the processes are defined. Like the entities of the ERD, the definition of the processes of the BPM must pass the proper criteria for quality of definition.

PROCESS DEFINITION CRITERIA

The criteria for quality of definition are almost identical with those found in data modeling:

- Does the process definition encompass *all* that it ought to encompass?
- Does the process definition encompass *only* what it ought to encompass?
- Is there a process in the scope of integration in the same dimension that is not covered by a definition?
- Has every term used in a definition been previously defined?
- Is there a process in the scope of integration that is covered by more than one definition?

- Are any two processes a form of a higher process (i.e., are processes abstracted properly)?
- Are the process definitions readable? Are they useful as a form of communication?
- Does the user concur with and understand the process definition?

As an example of the refining of a definition of a process, consider the following simple process definition: "claims processing is the consideration and/or settlement of a loss." The way the definition reads, *any* loss, whether or not insured, is covered, when in fact claims processing is for *insured* losses. So the definition now becomes: "claims processing is consideration and/or settlement of a formally insured loss."

Although this definition now covers most of claims processing, it does not apply to deductible buildup, because deductible buildup is not a formally insured loss. So the definition is refined: "claims processing is consideration and/or settlement of a formally insured loss or deductible."

The point is now made that certain terms are used that need to be defined more specifically. The term "settlement," for example, is one of those terms. A lawyer is called on to define a settlement: "relief of insured liability by means of a fiduciary surrendering of capital goods, monies, compensation, securities, or other bonds of indebtedness based on previously limited claims of insurability in accordance with terms of insurance fully in force at the time of loss incurrence."

Although the legal staff feels comfortable with the definition, no one else does. The user, data processing personnel, management, and the customer would all like the term "settlement" reduced to laymen's language. Someone suggests that a much more usable definition is: "Settlement is a fulfillment of insured coverage in the case of a loss."

As the processes and definitions are scrutinized for quality, the point is made that there may be overlap between "buying a policy" and "rate review." Rate review occurs whenever a policy is purchased or renewed. But given the BPM, rate review either does not occur for purchase or occurs redundantly as a part of purchasing a policy. Because the same activity is forced (artificially) to appear in two places in the BPM, it is suggested that the BPM be changed. Figure 7.6 illustrates the changed BPM, where rate review exists as a single activity. The reiteration of definitions and BPM structuring is a normal activity, as it is in the case of ERD definition.

The final step in the building of the BPM is to ensure that the user understand and concur with the definitions and the basic process model. Without user concurrence, the definition of the BPM becomes a "back-office" exercise and will ultimately carry little weight with the user.

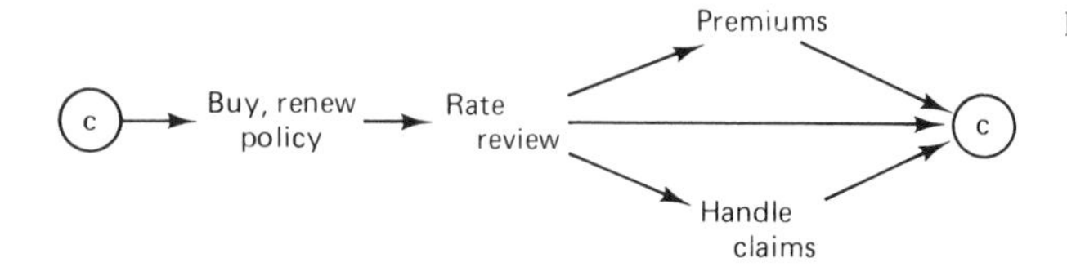

Figure 7.6 Revised basic process model.

DATA/PROCESS RELATIONSHIP

Even though data and processes are modeled abstractly and separately, ultimately they are embodied in a physical form and are interdependent and interrelated. To better envision the relationship of a process model and how it will ultimately operate in the real world, consider the next series of figures.

Figure 7.7 shows a data base with a predefined format of insurance data. The data shown is of three types: base data (name, address, etc.), rate data (factors influencing rates), and premium/payment data (where billing and payment data is recorded).

Figure 7.8 shows that the buy process has been activated. During the activation of the process, data is stored in the data base, as shown by filled-in areas. A policy number is established to identify the policy. After the buy process is finished, the policy number is passed to the rate review process, which activates the process as shown in Fig. 7.9. Now, the rate review process is activated and rates calculated based on existing data. The rates that are calculated are stored in the data base, as shown in Fig. 7.10.

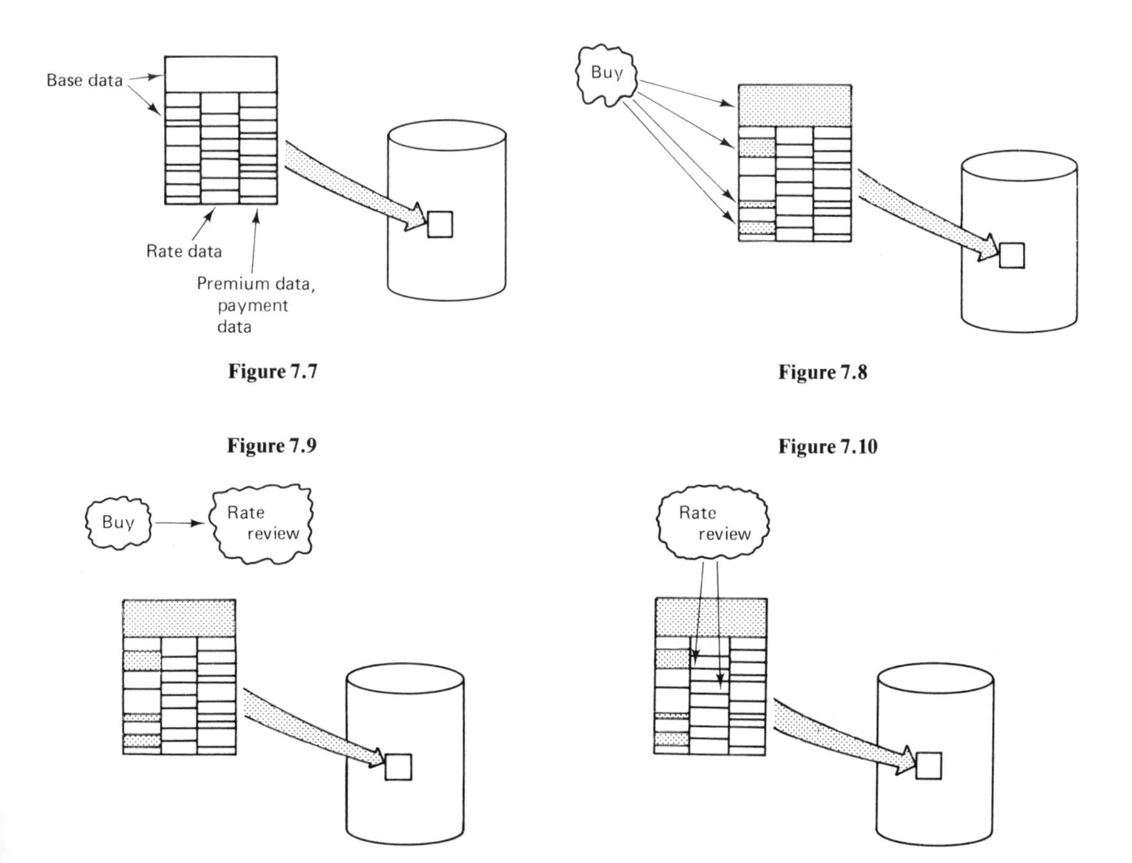

Figure 7.7

Figure 7.8

Figure 7.9

Figure 7.10

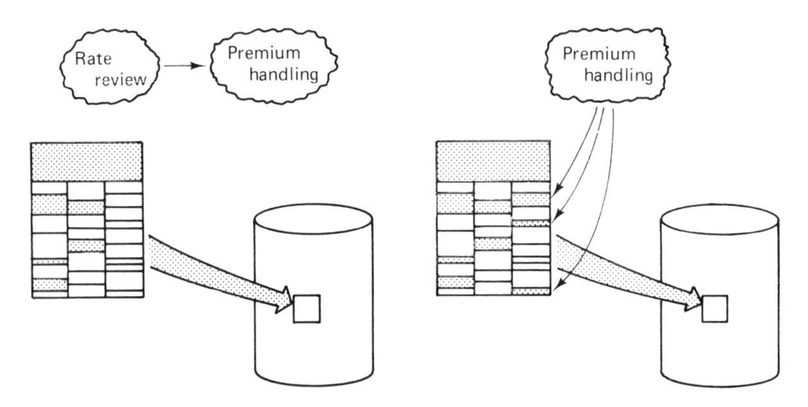

Figure 7.11 Figure 7.12

Upon termination of rate review processing, data is updated in the data base and control passes to the premium-handling process. The premium-handling process is activated by receipt of the policy number, which identifies the data for which a premium must be written, as shown by Fig. 7.11. Once activated, the premium-handling process calculates data and adds the data to the data bases, as shown in Fig. 7.12.

Upon completion of the premium processing, the cycle is ready to start anew. Note that data does not flow in an on-line, integrated environment as it does in a batch environment. The only data flow was that of policy number, and the only purpose there was to initiate a process, which could have been handled in another manner. Even though the batch practice of identifying data flow does not apply to the on-line, integrated environment, a similar design technique can be employed that is quite useful in documenting the activity of a process. The technique identifies what data must be available for the process to execute (which is analogous to the flow of data into the process). The data can be termed ''prerequisite'' data and ''transformation'' data. Prerequisite data is data that must be available for a process to go into execution. Transformation data is data that can be changed by the execution of a process. An occurrence of data may be both prerequisite and transformation at the same time.

One of the checks as to the quality of the design can be made centering around

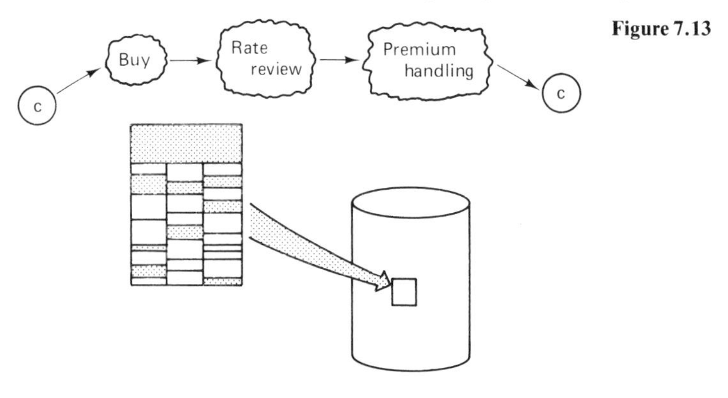

Figure 7.13

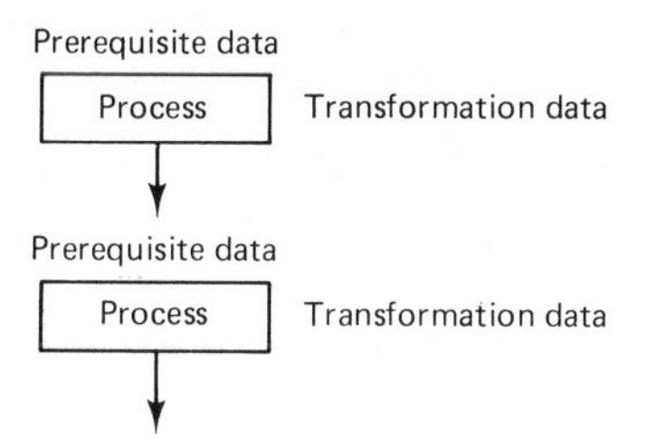

Figure 7.14

the fit of the prerequisite and transformation data, ensuring that a process does not execute until its prerequisite data is in place and the execution of the process results in the proper transformation of the data. Figure 7.13 shows a business cycle within the BPM and the data that relate to it. The prerequisite data for buying a policy might include policy issuance number, buyer name, address, agent number, policy type, and so on. All of the prerequisite data would also be transformation data in the case of a new policy. In the case of a renewal of a policy, only data that changed would be transformation data. The general form of showing prerequisite and transformation data is shown in Fig. 7.14. The actual usage of the form is shown in Fig. 7.15.

Figure 7.15

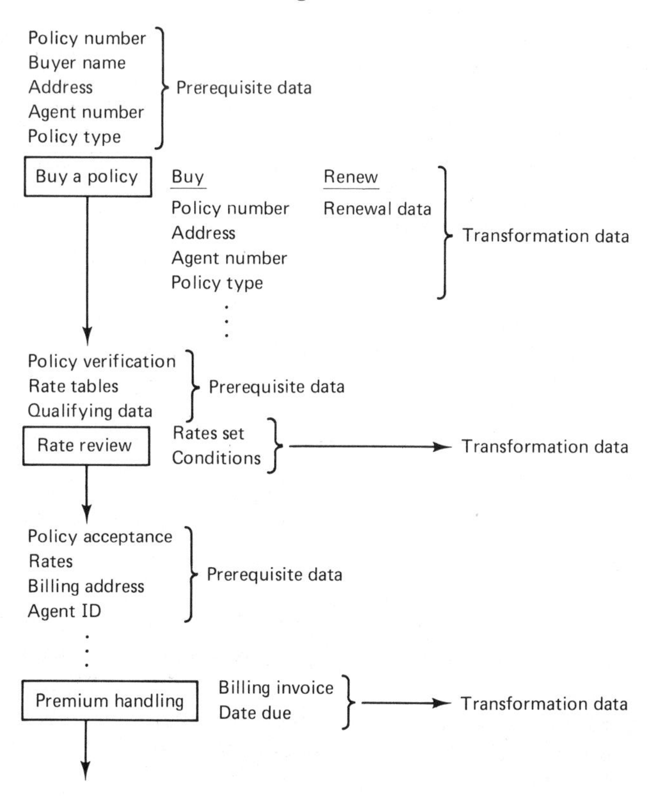

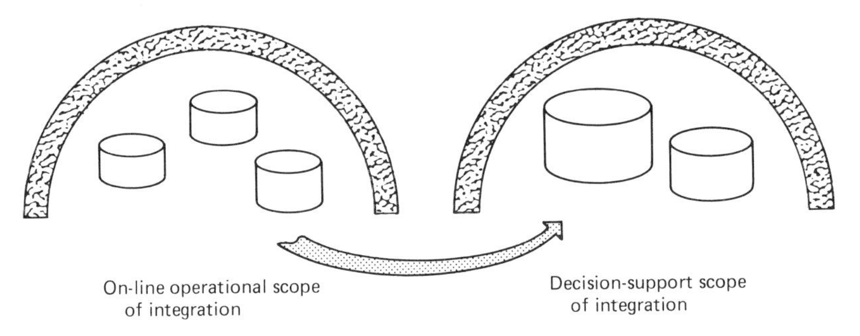

Figure 7.16 The flow of data across different scopes of integration is normal, necessary, and does not violate the sound practices of integration.

The prerequisite data for rate review would include established rates and policy verification. The transformation data include the rates set and any conditions that apply. The prerequisite data for premium handling include policy acceptance, rate determination, billing address, agent identification, and so on. The transformation data indicate that a billing has been made.

In a batch environment (which is commonly highly unintegrated) it is normal to pass much data (records, files, etc.) between processes or programs (hence the term "data flow diagrams"). But in a properly designed on-line, integrated environment there is no data flow, with one notable exception. The one exception occurs where there is a shift in modes of operation—from operational to decision support, from on-line to batch, and so on. Even in these cases, the data flow only across different modes of operation, which is *not* a violation of the sound practices of integration and system modeling. Within the same mode of operation there is no flow of data. Figure 7.16 illustrates a typical flow of data across ERDs and BPMs.

As a final note, the emphasis in building the BPM is on

- *Completeness:* Are all aspects of the dimension being modeled represented?
- *Simplicity:* Is the model concise? Understandable? Usable?
- *Organization:* Has the designer properly structured the flow?
- *Proper level of abstraction:* Are any two activities types of the same activity? If so, why aren't they combined?

IN SUMMARY

The BPM is to processes what the ERD is to data. The BPM, like the ERD, is extracted from the decomposition of the scope of integration. The cycle of processing is the basic form of the BPM. The processes in the BPM must be defined using the criteria for the quality of definition. The primary business-based BPM is constructed first, as is the primary business-based ERD.

With one exception, data does not flow in an integrated environment. The only flow that does occur is across modes of operation. The technique of analyzing prerequisite data and transformation data for each process is useful in determining the quality of design.

PROJECT STUDY

The analysis done by the IA team of ABC for process modeling coincides so nicely with the data modeling that the raw input from users is gathered once but is analyzed separately. The IA team of ABC meets separately from the data-modeling discussions to analyze process modeling for the scope of integration. The scope of integration for process modeling remains the same as the scope for data modeling—the financial systems of the ABC bank. Four major activities are outlined by the team for the basic process model (BPM), based on the results of the decomposition of the scope of integration. The four processes are: the acceptance of risk by the bank, the establishment of the terms of risk management between the bank and the customer, the transaction of activities against risk, and the periodic calculation of values of risk management. The term "risk" is used in the broadest banking sense. Risk can be either an extension of the bank's money or resources to an individual or enterprise in the expectation of repayment (e.g., a loan), or it can be the acceptance by the bank of money for stewardship (e.g., a savings deposit). The management of risk, then, is the highest level of abstraction within the scope of integration.

When a bank loans money, for example, it risks that it will have money returned with interest. When a bank accepts money for stewardship, it risks that it can keep the money safely and can make the money work to the point that the interest it pays the saver is able to be met. So at the highest level of abstraction, processing both loans and deposits are forms of managing risk.

The bank decides that there are four basic forms of risk (as far as normal financial services are concerned): loans, bankcard, savings, and DDA (checking). While the bank manages other types of risks (trust accounts, for example) the most basic risks (i.e., the risks most central to the business of ABC) are represented in the primary model. Also, more specific banking services (such as interest payment, interest collection, activity charges, etc.) can be classified as a common subset of the four basic risks that the IA team has selected.

So the primary business-based process model will include the basic BPM, as shown in Fig. 7.17. The basic business of the BPM cycle is depicted in the figure. Nearly all cycle activity is represented by a closed loop between the risk transaction activities and the calculation activities. The entry process—the acceptance of risk and the establishment of terms of risk management—is executed usually only once in the life cycle of an account. The periodic calculation of risk management values includes such activities as monthly interest calculations for savings and DDA, monthly statementing, overdue account analysis, and so on.

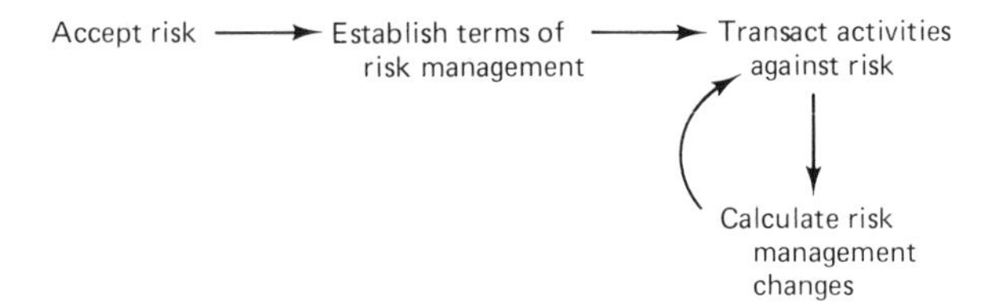

Figure 7.17

After the basic cycle is depicted, the IA team uses the criteria for quality of definition to create primary business-based definitions for the processes in the cycle. As an example of the process of defining a process, the IA team begins with a definition of "transaction against risk."

The first definition proposed is: "Transaction against risk is payment to an account." But it is brought up that withdrawals are as valid a form of changing an accounts balance as are payments to the account. So the definition now becomes: "Transaction against risk is the changing of an account balance."

However, this definition would legitimize activities outside the service agreement, such as overdrawing a DDA account or overcharging a bankcard account. So the definition now becomes: "Transaction against risk is changing of an account balance in accordance with the terms of the governing service agreement." Although this definition expresses most of what ABC means by transaction against risk, someone brings up the fact that monthly interest calculations fit the definition, whereas the IA team would like to include only customer-initiated activities in the definition. So the definition is refined one more time: "Transaction against risk is the changing of an account balance in accordance with the terms governing the service agreement that is initiated by a customer."

This final definition covers such things as loan payments, DDA deposits, savings withdrawals, and bankcard charges. It does not cover monthly statementing, monthly activity charging, monthly interest charges, and so on. In short, the process definition meets the criteria for quality of definition.

Once the definitions for the BPM are created, the IA ensures that the BPM has no activities that are types of a higher activity, and that the cycle of the BPM initiates the actual business cycle of the user.

1. List 10 activities that fit within the scope of integration and within the basic process model (BPM). List 10 activities that fit within the scope of integration and are not within the BPM.
2. What determines what should/should not be in the BPM?
3. What would have happened if the IA team had not defined the cycle around risk, but around lower forms of abstraction, such as loans, savings, DDA, and so on?

4. Does trust account management fit in with the BPM? Do money-market accounts?
5. Should prerequisite and transformation data be identified for the BPM? If so, identify the data.
6. When should the BPM be defined in relation to data-modeling activities?
7. Create definitions for the processes in the BPM other than "transaction against risk."

EXERCISES

1. (a) To achieve an integrated, on-line environment, it is necessary to model *both* data and processes. Why?
(b) Should either one be emphasized over the other? Why not?
(c) What happens if one is emphasized almost exclusively?

2. What are the issues of integration? Show how these issues ultimately boil down to money.

3. (a) What is process abstraction?
(b) Why should processes be abstracted?
(c) What happens if processes are not abstracted?
(d) Is integration possible in light of processes that have not been abstracted?

4. (a) Describe the components that make up a basic process model.
(b) Why are there very few components?
(c) Suggest three or four other components that might be useful.

5. Create a BPM for
(a) A bank
(b) An insurance company
(c) A manufacturing plant
(d) A distribution and marketing company

6. Using the criteria for quality of definitions, define the following:
(a) Order process
(b) Bill of materials process
(c) Accounts receivable process
(d) Premium billing process

7. Identify the prerequisite data and transformed data for the BPMs created in Exercise 5.

8. (a) Can data flow in an integrated environment? Should data flow?
(b) Why is data flow normal in the batch environment and an exception in the integrated environment?

8

Function Modeling

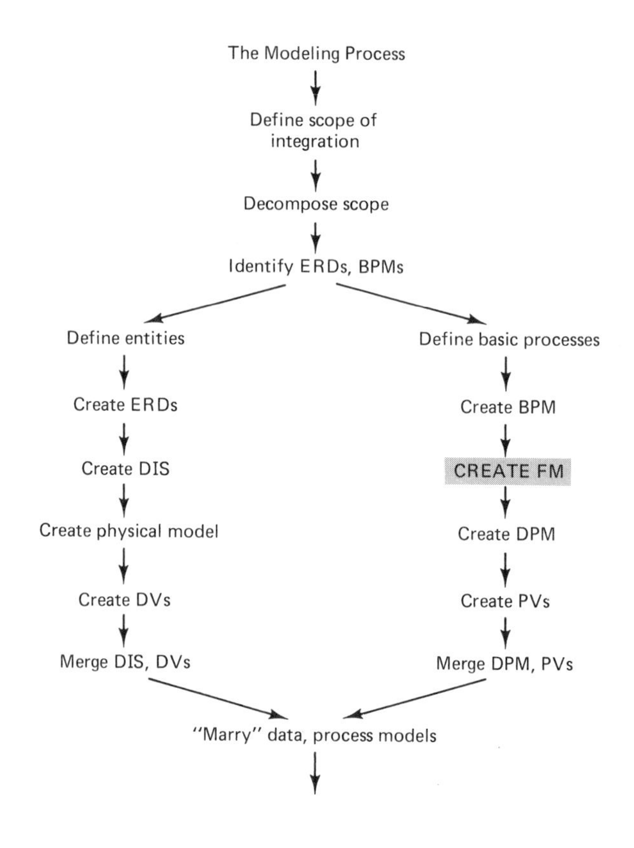

The purpose of the basic process model (BPM) is to outline *all* the functions that make up a dimension, to identify them, and to define the cycle in which they operate. The BPM is prepared at a very basic level. The quality of the BPM—its completeness, representation, and so on—all depend on the business of the organization (i.e., the business of the organization determines what is correct and what is incorrect). The cycle of the BPM is such that the end of one cycle starts another. For other cycles at a lower level, this is not necessarily (in fact, is usually *not*) the case. The next level of detail below, a BPM is a function model (FM). In its simplest form a function model is merely an expansion of each activity found in the BPM, but there are usually many more refinements in the FM. The expansion of the BPM into function models is analogous to the expansion of each of the entities of the ERD into a DIS.

The function model expands each activity of the process cycle into a series of ordered events (thus creating a smaller cycle). These FM activities (the FM cycle) are unique and, when combined with other FMs, completely define the BPM. As an example of an expansion of an activity from the BPM to the function model, consider Fig. 8.1. In this example the basic process activity of claims handling has been broken down into the major activities (at the FM level) of receiving claims, data verification, coverage calculation, and check issuance. The major activities of claims handling are represented at the functional level. They represent *all* the activities that are a part of claims processing, and they are sequenced in the order in which they are normally executed, creating their own functional cycle.

The process is set in motion (i.e., the cycle is activated) when it is "entered" (i.e., a claim is received that must be processed). The process is completed by "exiting." Upon exiting, control is not returned to the beginning of the cycle. There are a few exceptional circumstances that will occur where the normal exit is not used. However, for the vast majority of the processing, the exit represents the final point of control of the process.

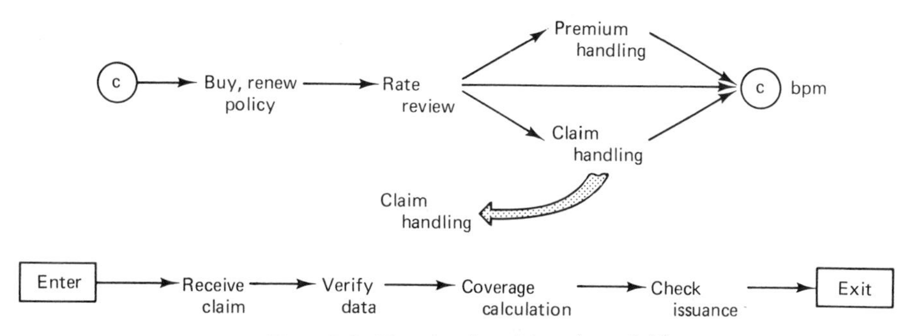

Figure 8.1 Functional model: major activities.

MAJOR ACTIVITIES EXPANSION

Once the major activities have been identified and ordered, the next step is to outline lower-level activities that support the major activities. Such a breakdown is shown in Fig. 8.2. This figure shows the activities as a mere listing of activities (without any

Receive claim	Verify data	Coverage calculation	Check issuance
Post receipt	Verify policy	Other coverage?	Record amount by policy, agency, type of coverage
Notify agent	Verify agent	Liability maximum?	Issue check(s), depending on size, location, etc.
Classify claims type	Recorded data check	Fault/no fault?	Determine shared party percentages
Locate policy	Reasonability check	Act of God?	
	Past claims check	Basic act coverage?	
	Past rate reviews	Environmental factors?	
	Related claims	Pending litigation?	
	Coverage data check	Shared responsibility?	
		Automated calculation?	
		Case review (if beyond reasonableness)	

Figure 8.2 Subactivities of claim handling.

implied meaning as to order of occurrence or other details). The purpose of this first cut at fleshing out the details of the major activities within the FM concentrates on the identification and completeness of activities. All the major activities and subactivities of the basic process should be identified. All the different types of detailed activities that comprise an FM activity are identified. After all activities are identified, the next step is to refine those activities according to

- Order of occurrence
- Logic that alters flow
- Data that is used by the process

In essence, each cycle that is appropriate to a lower level of processing is identified and organized.

ARRANGING THE ACTIVITIES

After the activities have been identified, they must be arranged in the order in which they will be executed. A work flow is established, much like the work flow through the BPM, except that this work flow is through the major activity of a single basic process. The prerequisite and transformation data may or may not be shown. Such a work flow, shown in Fig. 8.3, has several points of interest. The first point occurs when litigation is pending, and an abnormal exit is taken. This means that the activities for claims processing in the face of pending litigation are not defined as a regular part of the claims processing flow. This instance may or may not be defined at all within the definition of the normal insurance cycle.

The second point of interest is that the order of activities is arranged in a filtered manner (i.e., those activities and decisions that would eliminate a claim from processing are done first). The third point is that there is a linear flow from the entry to the exit (i.e., the start of the cycle to the end of the cycle). If a branch is taken, it results in a rather immediate return to the central path of activities. Only an exit that eliminates a claim does not follow this pattern.

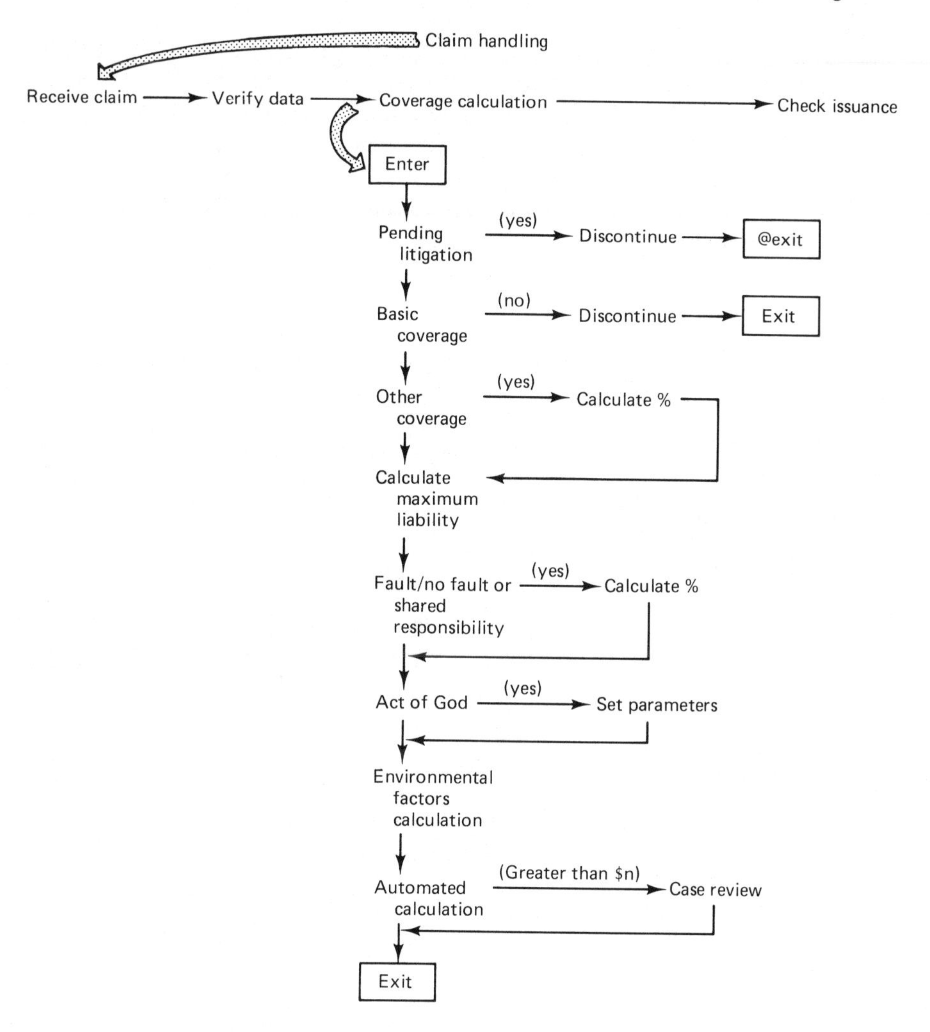

Figure 8.3 Flow of activities of coverage calculation.

A fourth point is that prerequisite data must be in place when it comes time for the execution of a process. Another point is that the activity is designed to process a mainstream of flow (i.e., most claims will pass through the logic of processing without ever having veered from the normal flow). And yet another point is that by the time the coverage calculation is reached, there is no question as to such basic issues as: Does the claimee have a policy? Is it in force? and so on. Such basic questions have already been asked and successfully answered (i.e., the prerequisite data has previously been satisfied).

The flow of activities essentially satisfies the first two criteria for the refinement of function: the definition of the order of occurrence of activities and determination of any logic that alters the flow. But there is a third criterion that is needed to complete this phase of process modeling, and that criterion is a mapping of the data that is used by the process. What data is needed at what point in the process? Such a mapping is shown in Fig. 8.4, where, at each point in the process, the type of data needed is identified. Note that the identification is *not detailed*—only the general type of data and where it is likely to reside is indicated. The reason for the ambiguity is that the corresponding data model may or may not be sufficiently detailed at this time.

Another way of indicating what data is associated with a process is shown in Fig. 8.5. A point of interest is that data does not flow from one activity to the next (as has previously been discussed). Each activity accesses, manipulates, and updates data in a central location and then passes control to the next activity. This freedom from data flow has previously been noted at the BPM level. Were data to exist redundantly, flow would be necessary.

Figure 8.4 Flow of activities of coverage calculation.

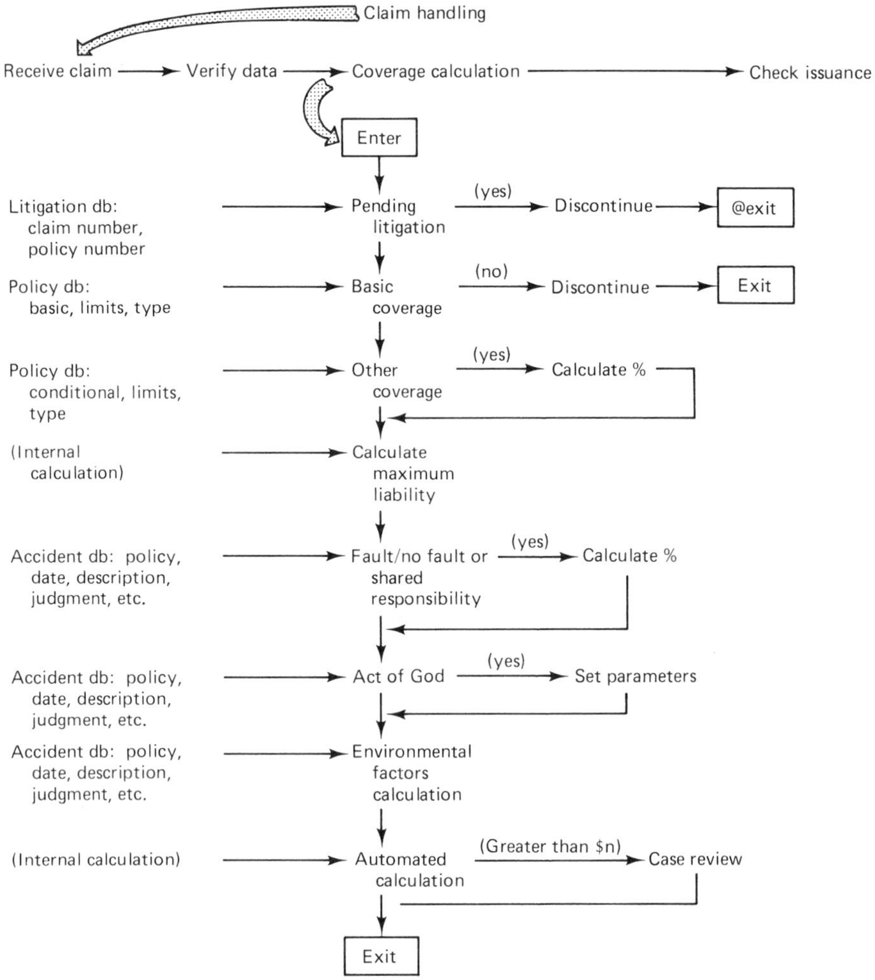

Claim handling

Receive claim → Verify data → Coverage calculation → Check issuance

Enter

Litigation db: claim number, policy number → Pending litigation —(yes)→ Discontinue → @exit

Policy db: basic, limits, type → Basic coverage —(no)→ Discontinue → Exit

Policy db: conditional, limits, type → Other coverage —(yes)→ Calculate %

(Internal calculation) → Calculate maximum liability

Accident db: policy, date, description, judgment, etc. → Fault/no fault or shared responsibility —(yes)→ Calculate %

Accident db: policy, date, description, judgment, etc. → Act of God —(yes)→ Set parameters

Accident db: policy, date, description, judgment, etc. → Environmental factors calculation

(Internal calculation) → Automated calculation —(Greater than $n)→ Case review

Exit

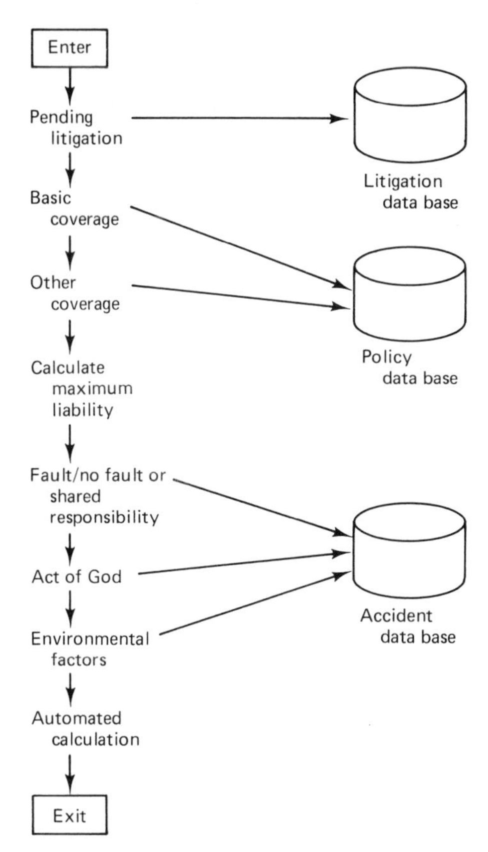

Figure 8.5 Claims coverage activities versus data accessed and/or updated.

FUNCTIONAL MODELING

There are many ways to decompose a major functional activity, and each of the ways has its own set of merits. There are a few standard practices that greatly enhance the chances of making the decomposition more effective. Some of these practices are:

Linearity of Flow

The flow of activities within the processes should be as linear as possible. Any branches should either go to exits or back to the main flow shortly. It is confusing to have more than one primary flow established. The flow should represent the mainstream of processing and should contain activities that are unique (i.e., not found elsewhere). As an example of improper flow construction, refer to Fig. 8.6. The rate review is divided into reviews for new policies and old policies. Although there are differences in the flow of activities, there is much redundancy between the two policy reviews. The proper way to handle the reviews is to create a single activity

Rate review

New policy	Policy renewal
Get data from policy application	Get data from existing policy
Get data from actuarial base	Get data from actuarial base
Verify data (reasonableness, cross-checking, examination, etc.)	Get claims data
Categorize applicant	Get agents assessment
Weight categories	Get record of insurability
Calculate risk based on common tables	Verify data (reasonableness, cross-checking, examination, etc.)
Assign rates based on risk	Categorize applicant
	Weight categories
	Calculate risk based on common tables
	Assign rates based on risk

Figure 8.6 Improper activity flow; there are two main streams that contain redundant activity.

Rate review

Get data from policy: new – application
old – existing policy

Get data from actuarial base
(old): Get claims data
(old): Get agent's assessment
(old): Get record of insurability
Verify data (reasonableness, cross-checking, examination, etc.)
Categorize applicant
Weight categories
Calculate risk based on common tables
Assign rates based on risk

Figure 8.7 Proper activity flow; there is a single mainstream and no redundant activity.

flow with nonredundant activities. (This is in accordance with the concept of abstraction, where common processes are consolidated.)

The flow makes allowances for unique activities where they occur. The proper flow is shown in Fig. 8.7.

Unity of Purpose

Each activity description should describe nothing more than the activity (i.e., extraneous, unrelated, and tangential activity descriptions need to be described separately, in their own functional model). For example, consider the activity flow for rate review when it comes time to describe checks of reasonableness, as shown in Fig. 8.8. The business activities described are a part of the general activity of the business of rate review, but serve their own separate and smaller purpose. To avoid complexity and to reduce the size of the description, the reasonability analysis is set off as its own set of activities, as shown in Fig. 8.9. Of course, a reference is made to the fact that a reasonability check will be made in the rate review, but the reference will be concise, not interrupting the train of thought of the person trying to understand rate review processing.

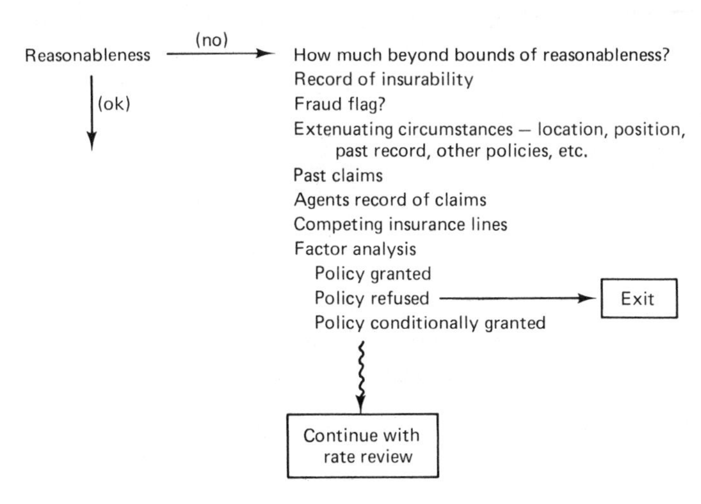

Figure 8.8

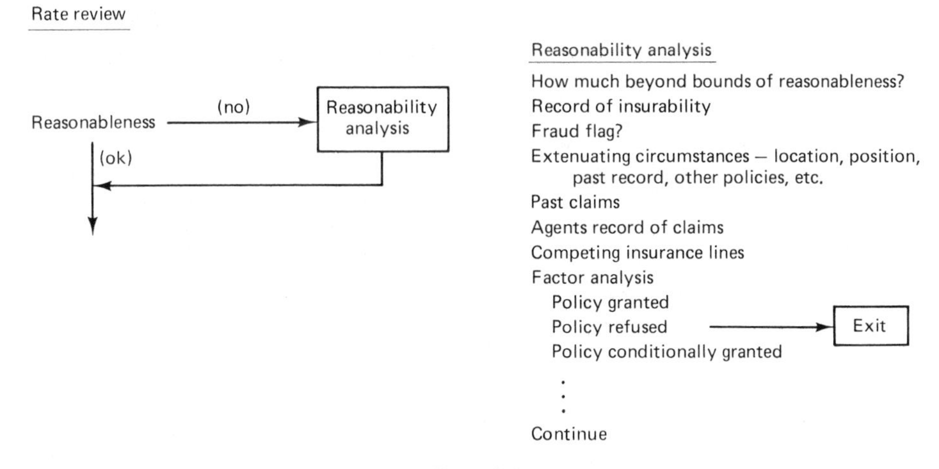

Figure 8.9

Separation of different types of processing has the effect of capsulizing the activity, which greatly reduces the volume of documentation needed to describe an activity. Without the ability to capsulize and subset, activity descriptions become very large and because there are many activities within a major activity, difficult to follow.

High Level of Detail

Even though functional modeling represents a decomposition of major system activities (and as such reduces processes to a lower level of detail), nevertheless, the level of detail is not as low as is needed to write program specifications from. For example, functional modeling does not (often) detail activity about different types of processes that have been abstracted together. For example, the functional model says very little or nothing about auto, home, fire, commercial, or any other type of insurance. Each of these is a type of insurance and is taken into consideration by the higher form of processing. If there is a mention of type made, the mention is vague and has little or no detail. Figure 8.10 shows part of rate review. The system architect assumes that later analysis will lead to the differences in processing between the different types of insurance as far as calculation of risk is concerned. This is shown on the left side of Fig. 8.10. If it is felt that further notice is worthwhile, the activity flow may be expanded as shown on the right side of Fig. 8.10. But even when auto, home, and so on, are mentioned, the reference is nondetailed. The combination of processing at this level assures that commonality is recognized.

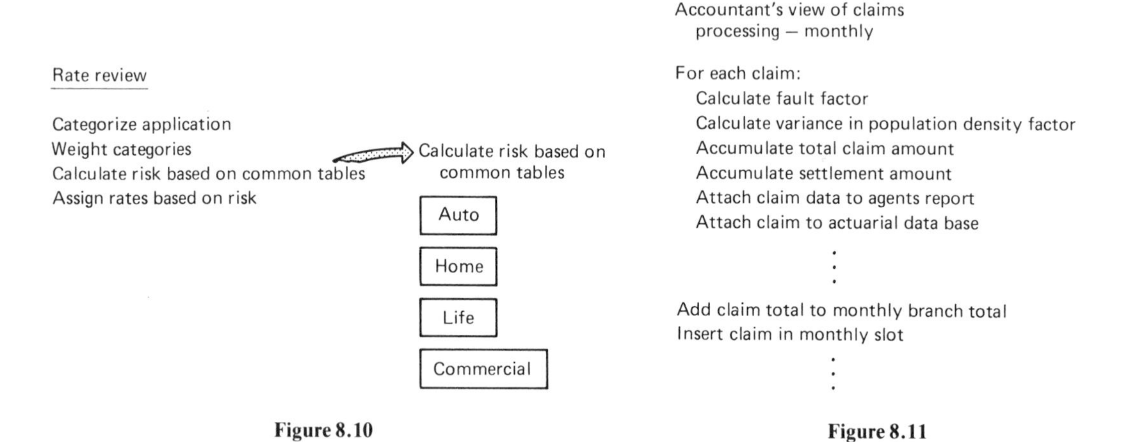

Figure 8.10

Figure 8.11

OTHER CONSTRUCTS

There is another useful construct that has not been shown thus far. The FOR ... construct allows a process to reiterate certain activities and still exist in the mainstream description. As an example, consider Fig. 8.11. In this figure an accountant's end-of-month processing is represented and a certain number of claims are accumulated each month. At the end of the month, each claim must be processed. Figure 8.11 shows part of the processing by means of a FOR... construct.

"TYPE OF" EXTENSION OF THE FM

Once the basic flow of the FM is decided, there is a further step that can be taken. Whereas the basic flow of the FM shows how the FM's activities are to be organized, it does not show much detail about the activities. In particular, it does not account for the different types of activities that may comprise each FM's activity.

Figure 8.12 shows how the FM activities can be expanded. In the figure basic coverage is broken out by fire, home, and life. If there were other types of insurance in the system, they would be included. The breakout of types uses the "type of"

Figure 8.12

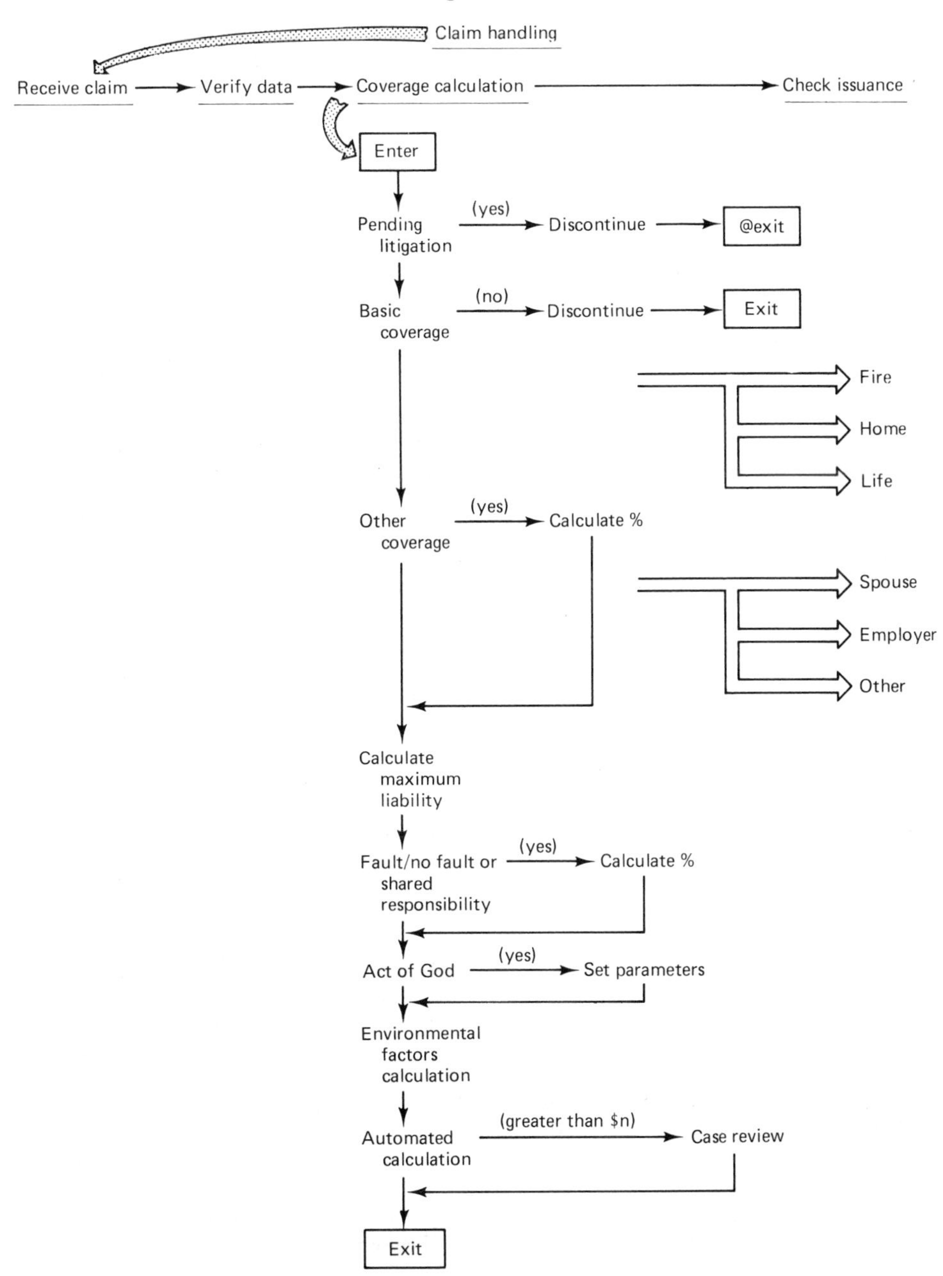

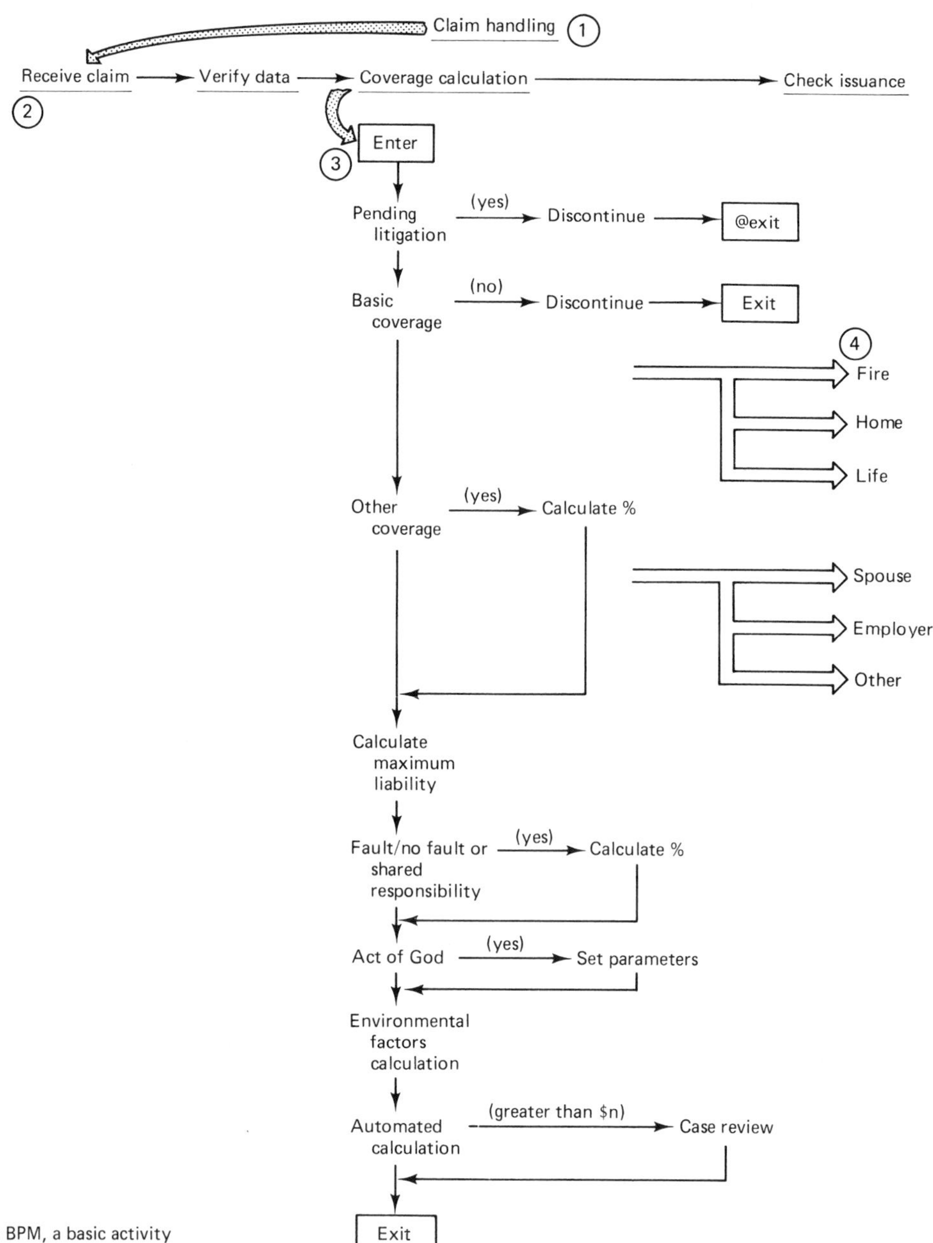

1. From the BPM, a basic activity
2. The FM created from the activity in the BPM
3. The process outline of an activity in the FM
4. The various types of activities that make up part of the process outline (using "is a type of" notation)

Figure 8.13

notation. The diagram reads: "Is there basic fire coverage? If no, discontinue." "Is there basic home coverage? If no, discontinue." "Is there basic life coverage? If no, discontinue." The same sort of breakdown by type occurs for other coverage. If the insurance has extra spousal coverage, employer coverage, or other coverage, a percentage is calculated.

Note that the actual processing detail for "type of" processes is not included here. It will be included at a lower level of detail.

IN SUMMARY

The BPM is created from the decomposed scope of integration. The BPM is the cycle of business most essential to the organization. The BPM represents the primary business processing of the organization. The functional model is created from the BPM and represents an expansion of the BPM. The functional model represents the flows of activities of each major processing activity and as such represents its own business cycle. There is a single flow of activities that are modeled functionally which corresponds with the major processing flow of the activities being modeled. The activities in the functional model are nonredundant and can be capsulized to achieve conciseness and clarity.

Figure 8.13 shows the general flow of analytical activities. First, the BPM is created and is organized into the most basic activities of the organization. Next, each basic activity is broken into a lower, functional level. At this level different lower-level activities are identified. These functional activities are further divided into a process flow. If appropriate, activities in the process flow are divided into their respective types.

PROJECT STUDY

After the IA team has created the BPM and has created definitions for each basic business process, each process is then expanded into a more specific form—an activity at the functional model (FM) level. There will be four types of processes modeled at the FM level: loans, DDA, savings, and bankcard. The processing cycles for each of these types is shown in Fig. 8.14.

Each type of process does, in fact, correspond to a higher form of the BPM, even though at the FM level there are differences between the different types. Some of the differences are outlined in Fig. 8.15. These differences at the FM level are the result of modeling different dimensions of processes. For example, the stop payment of a check is not part of the primary business-based view of a process simply because stop payments occur so infrequently (however important they are when they do occur!). Of course, eventually, the stop payments dimension must be accounted for by the process model, but not as a part of the functional processing model.

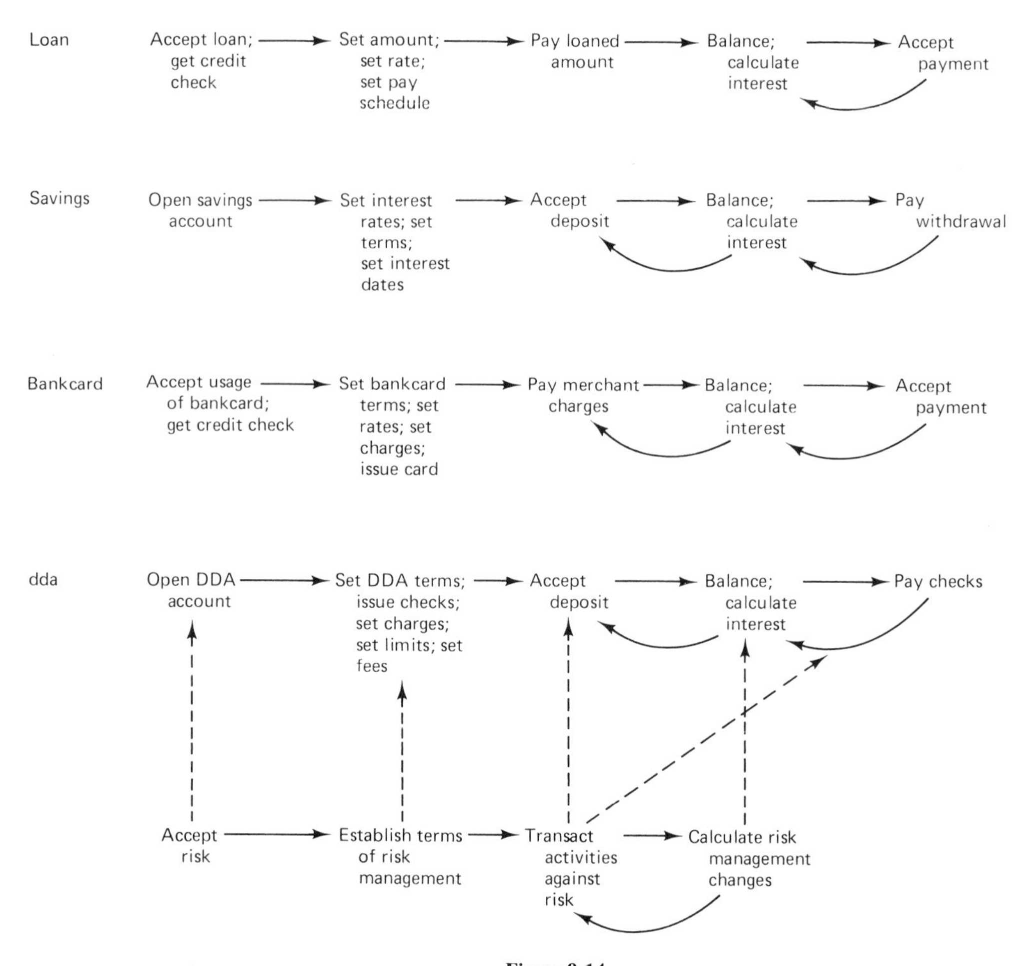

Figure 8.14

Figure 8.15

Missing processes:

DDA:	Stop payment, overdraft protection, overdraft management, account linkage, closing account, statementing, adjustments
Bankcard:	Card closing, disputed claims, nonpayment, limit adjustment, charge adjustment, statementing, account linkage, associate account, on-line balance query
Savings:	Account closing, account linkage, associate accounts, account transfer, "pledged" savings
Loans:	Account closing, loan renewal, foreclosure, account linkage, associate accounts

These processes are a normal part of the life of a bank-managed risk, but they do not belong in the primary business-based model. Instead, they represent a different dimension(s).

Raising a loan limit is not a part of the primary business-based FM because it represents a reiteration of the risk acceptance process—the initialization of the process. Also, the raising of a loan limit occurs in a small percentage of the processes that are executed. (*Note:* A pledged savings account is one where balance cannot drop below a certain amount and is used as collateral for other accounts, usually loans). All of these nonprimary business-based dimensions will be accounted for in the final process model, but the minor differences they represent only create an obstacle to the consideration of the mainstream activities.

The IA team takes each BPM process and identifies the corresponding fm process, as seen in Fig. 8.16. As shown, the FM activities can be divided into two classes: payments and credits. Both payments and credits can in turn be abstracted to a higher form—the changing of balances.

1. For each BPM process, identify the corresponding FM process.
2. For similar FM processes, what common functions are there? What distinct functions are there?
3. Can payments and credits be abstracted to the same level? Why or why not?
4. Who should do the separation of processes from the BPM to the FM?
5. What data modeling activities should be occurring simultaneously with FM activities?

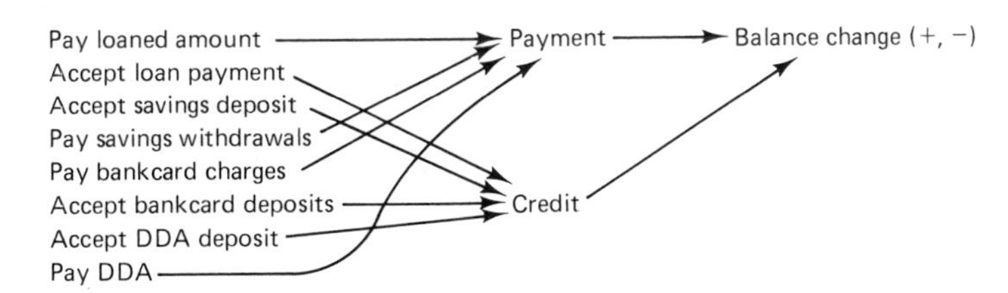

Figure 8.16 Activities transacted against risk.

EXERCISES

1. (a) Do all BPMs need to be broken down functionally?
   (b) Is there an implied order for the breaking down of BPMs?
2. What happens when a BPM is broken down functionally and shares common functions with other BPMs?
3. Create a functional model for the BPMs created in Exercises 5, Chapter 7.
4. Can program specifications be written from a functional model of the data? Why? Why not?
5. (a) Create a functional model that uses the FOR. . .construct.
   (b) Suggest ways to document the FOR. . .construct for greatest clarity.

9

Process Modeling: Detailed Process Model

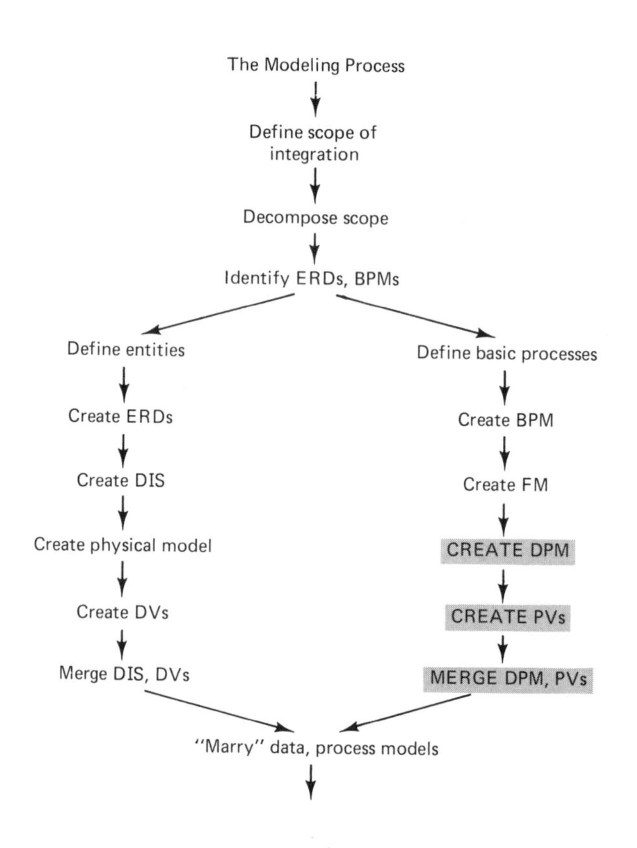

The detailed process model (DPM) is constructed from the processes found at the functional modeling level in much the same way that the functional model is derived from the basic process model. As in the case of basic process model to function model, the detailed process model is a representation of a process (the activities and the ensuing cycle) at a more detailed level. From the BPM program, specifications can be created. The relationship between the levels of modeling is shown in Fig. 9.1, where the basic insurance cycle (for all policies, in all locations) is outlined at the BPM level. At the functional model level, one of the major processes—claims handling—is broken down. Of course, for a complete functional model, *all* major processes would be broken down. At the detailed process model level, the function of individual policy coverage is further defined (in the example shown).

At the detailed process model level, two types of processing are defined: processes common to more than one function and processes that are unique. The for-

Figure 9.1

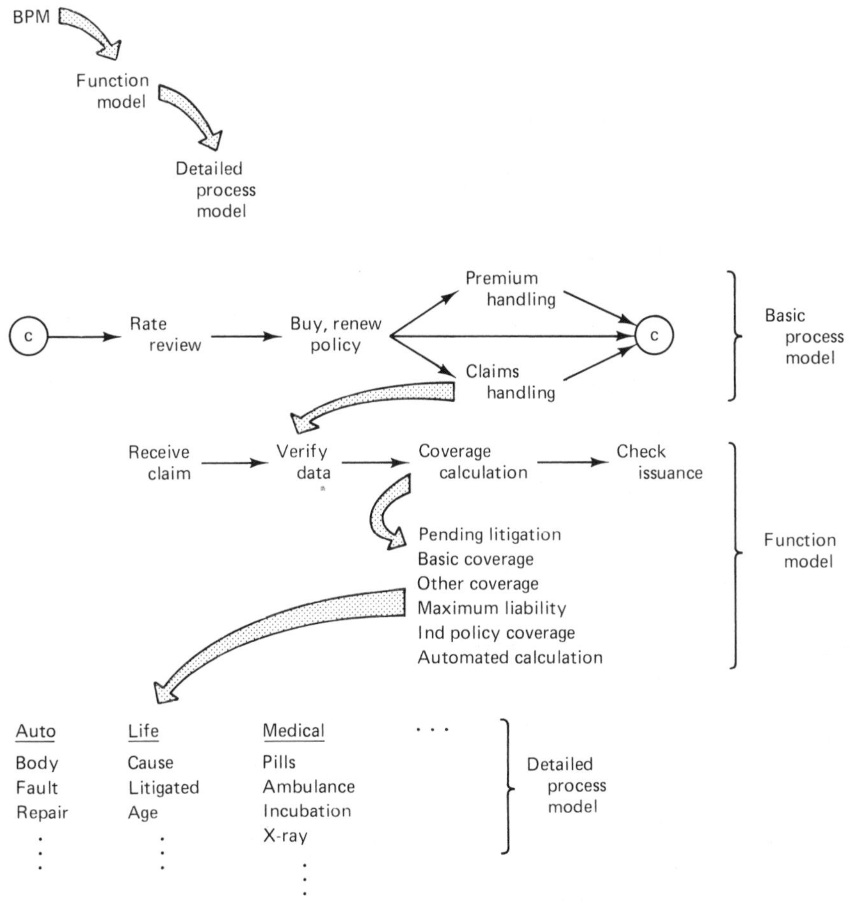

mat and techniques of describing the process model are the same as those at the functional level. As an example of the decomposition into detailed process model form, consider Fig. 9.2, where the details of the calculation of maximum liability are being determined. The first step is to break liability down by types of insurance: auto, life, medical, and so on. This is done at a gross level in the first pass. It is noteworthy that

Figure 9.2

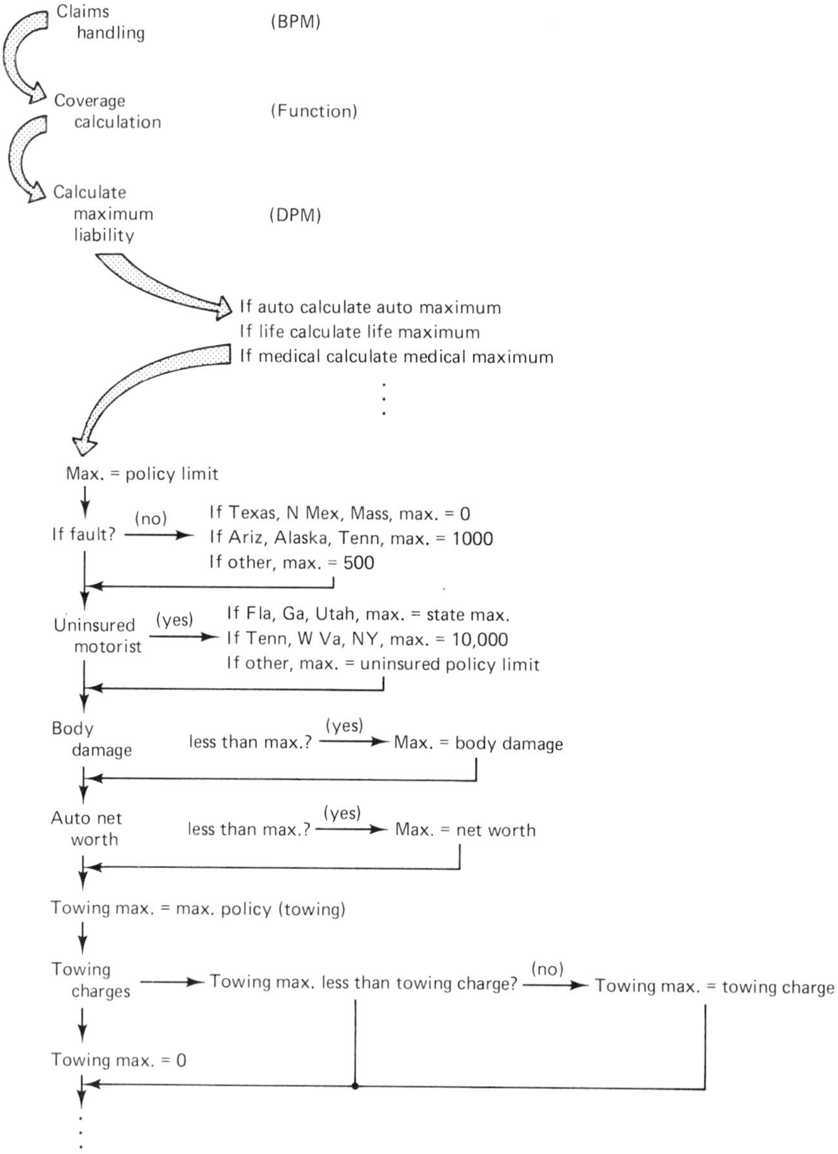

Claims handling (BPM)

Coverage calculation (Function)

Calculate maximum liability (DPM)

If auto calculate auto maximum
If life calculate life maximum
If medical calculate medical maximum
⋮

Max. = policy limit

If fault? (no) → If Texas, N Mex, Mass, max. = 0
If Ariz, Alaska, Tenn, max. = 1000
If other, max. = 500

Uninsured motorist (yes) → If Fla, Ga, Utah, max. = state max.
If Tenn, W Va, NY, max. = 10,000
If other, max. = uninsured policy limit

Body damage less than max.? (yes) → Max. = body damage

Auto net worth less than max.? (yes) → Max. = net worth

Towing max. = max. policy (towing)

Towing charges → Towing max. less than towing charge? (no) → Towing max. = towing charge

Towing max. = 0

⋮

at this point the differences in processes begin to surface. Prior to this point of analysis, functions were discussed at a common or generalized level.

After the types of insurance are outlined, the logic that is used is detailed, as shown in the remainder of Fig. 9.2. The logic is linear (as it was at the functional level), and the same logic is not repeated elsewhere (e.g., whether a fault has been determined or not is considered in only one place). Also of interest is that individual state considerations are made, as seen in the fault and uninsured motorist logic.

It would have been possible to break down maximum liability calculation by state first, then by function. Such a breakdown would have had much redundant logic. For example, fault would have been analyzed in 50 different places (one for each state). With the decomposition of function that has been chosen, any given function is written once, and individual state considerations are made throughout the process breakdown. This means that as the insurance laws in Texas undergo massive changes, much logic (source code, documentation, etc.) must be searched. However, if process decomposition is done by state, it makes accommodating state changes easy but at the expense of basic calculations central to the business of the insurance company. This trade-off is illustrated in Fig. 9.3.

Figure 9.3

State organization/ business organization	Business organization/ state organization
Texas	Function a
Function a	Texas . . .
Function b	W Va . . .
Function c	Function b
N Mex	Texas . . .
Function d	Arkansas . . .
Utah	Oregon . . .
Function a	Illinois . . .
Function d	Function c
Function f	Texas . . .
Function g	New York . . .
Wyoming	Maine . . .
Function g	Georgia . . .
Colorado	Function d
Function e	N Mex . . .
Function f	Utah . . .
Function g	Kansas . . .
Arkansas	Missouri . . .
Function b	
Function d	
. . .	
State processing must be changed in a single place; function processing must be changed in multiple places.	The code representing a function must be changed in a single place; the code representing a state must be changed in multiple places

After the processes have been decomposed, the next step is to identify what data will be necessary to satisfy the function. Identification is done at a fairly detailed level, outlining the general collection of data and the specific items that are necessary at each step in the process, as shown in Fig. 9.4. Of interest in this figure is that state limitations on fault and uninsurance are found in two places, in the process definition itself and in the state data. When limits are defined in the process, change becomes difficult when a given state legislation makes changes. When data is defined externally to the process, changes can be made by a state legislature without affecting the process definition. By the same token, when specific values are defined in a process, no input or output activity is required, thus making the process very efficient to execute. When data is defined externally, input must occur for the process to execute, thus slowing down the execution of the process. The rightness or wrongness of whether data is to be hard-coded depends on:

- The frequency of the change of the value
- The frequency of execution of the process
- The speed of execution desired

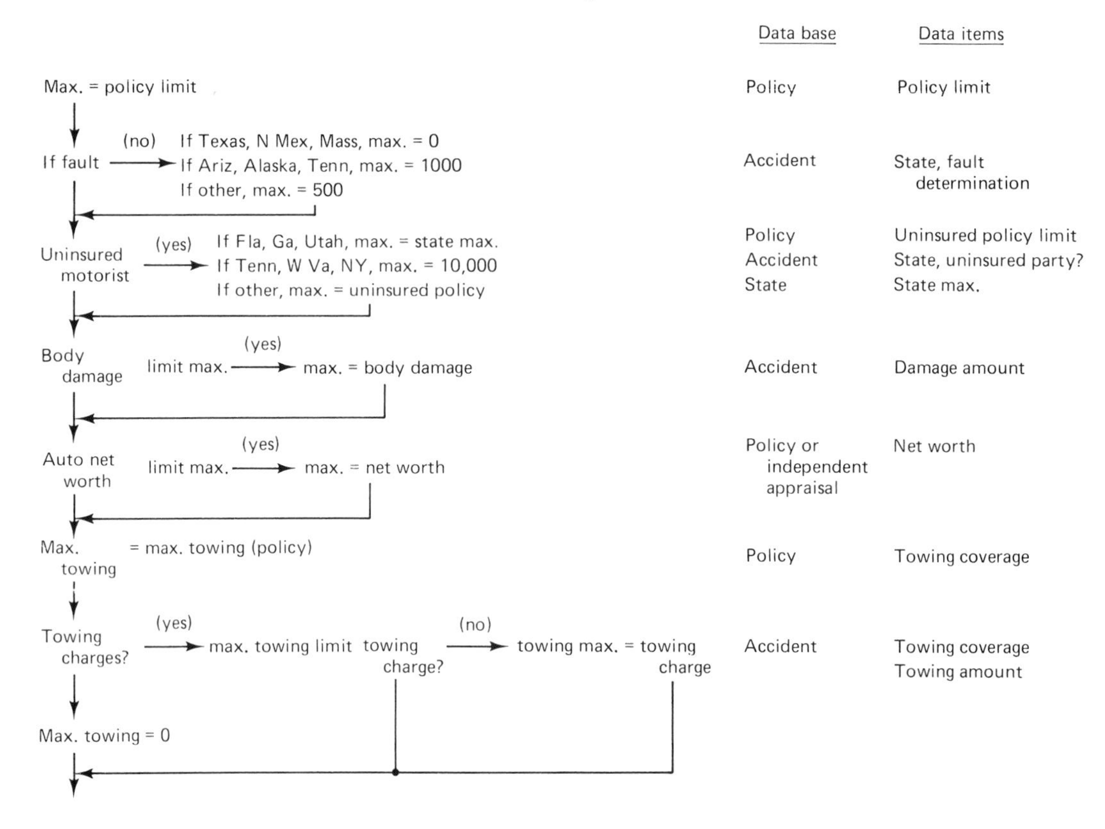

Figure 9.4

REDUNDANT PROCESS DEFINITION

As a designer proceeds top down from the BPM to the DPM, one very specific goal is to remove all redundant process definition from the system. There are *many* valid reasons for removing process redundancy at the logical system design level. Some of them are:

- Redundant process definitions require multiple development efforts.
- Redundant process definitions require multiple maintenance efforts.
- Redundant process definitions often require intricate synchronization of processing.

However, as the designer proceeds downward in top-down design it becomes apparent that at the implementation level there is a very real need for process redundancy. (The same phenomenon is observed in data—data is unified at the ERD level and structurally redundant at the physical level). It is quite normal for a process to be executed in multiple places, but it should be defined only once. For example, exponential smoothing of inventory usage in manufacturing can occur in many different places in manufacturing systems. But the source code for exponential smoothing should occur only once. The source code is then invoked as a subroutine where needed. There are two basic types of redundant processes:

- Utilitarian
- Multiple-occurring functions

As an example of a redundant utilitarian process, consider a bank that must calculate realized interest rate from a simple annual rate. Such a calculation must be made for loans, advertising, marketing, and in many places. Even though the utilitarian function is found in many and varied functional areas, there is no need to redefine and reconstruct the utilitarian function everywhere it is needed. As long as the *total* system requirements for such a function are defined at the outset, the utilitarian function can be built once at the source code level and used interchangeably as a subroutine in many places.

The other type of redundant process that occurs is that which occurs in multiple places. Looking at a taxonomy of the functions of a system in environments, such as insurance, banking, manufacturing, etc., it is apparent that some major functions are repeated in many places. One of the principal goals of building a business-based model is to identify those functions and consolidate them logically, even though at the implementation level the functions will exist in many places. ***As long as the function is logically defined in a single place, it can be constructed and maintained once and exist physically in multiple places. But if a function is not defined logically in a consolidated manner, it cannot be constructed and maintained in a single place.***

This very important concept is best understood in terms of some examples. Consider the insurance policy issuance system depicted in Fig. 9.5. The policies can be issued locally (at an office), at a regional office (at a branch), or at regional headquarters. Most customers can be satisfied at the local level, but some customers (large accounts or very specialized policies) must be handled elsewhere. The process required to issue a policy is the same for all levels, the only difference being that headquarters can perform certain functions that the local office cannot. It does not make sense to logically define a separate policy issuance for each level. But the actual program that issues policies will be different from one level to the next. The policy issuance program for the office will run on a personal computer or a minicomputer. The policy issuance program for the branch will run on a minicomputer, while head-

Figure 9.5 Detained processing model: example of the usefulness of functional integration.

Headquarters policy system	Branch policy system	Office policy system
Mainframe 5% of customers Customized policies	Mini 15% of customers Group policies, up-scale policies, unusual limits A subset of headquarters policy system	Personal computer 80% of customers Normal coverage, normal limits Subset of branch policy system

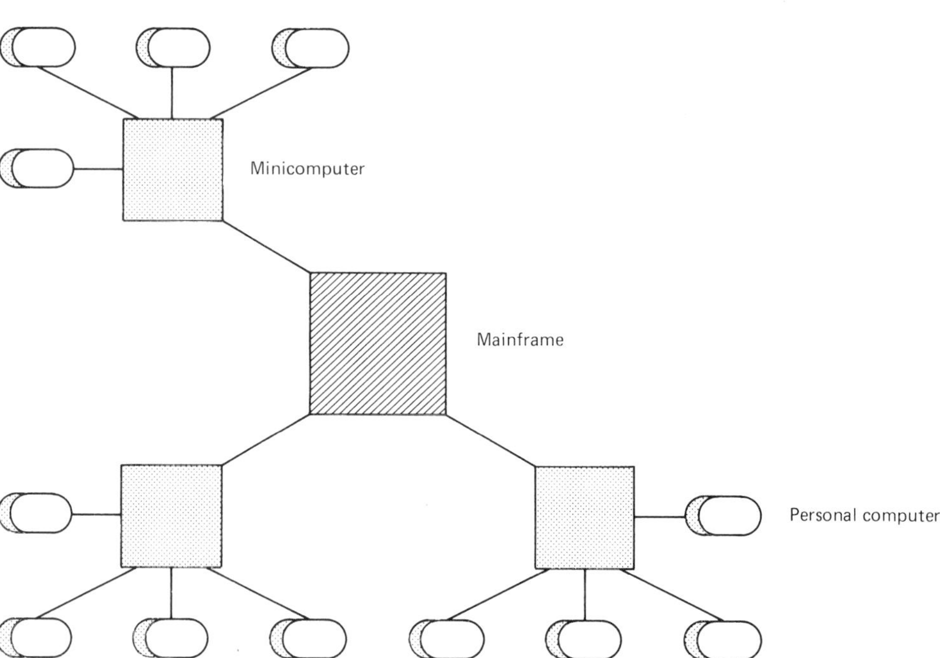

quarters runs their work on a mainframe. The relationship is shown in the lower part of Fig. 9.5.

The logical design of the system shows no differences in policy issuance at the logical, or conceptual, level. At the logical level there is no redundancy. But at the physical level (on the different computers) there is *much* redundancy. The programs that run on one personal computer are redundant with the programs that run on *all* other personal computers. The same holds true for the programs that run on the minicomputers. There may also be redundancy of code between the mini-, personal, and mainframe computers. So redundancy abounds at the physical level, whereas there is *no* redundancy at the logical level.

OTHER PHYSICAL DIVISIONS

As another example of logical unity and physical redundancy, consider the claims processing system shown in Fig. 9.6. The claims processing system is distributed to four physically separate locations, and each location summarizes data periodically

Figure 9.6

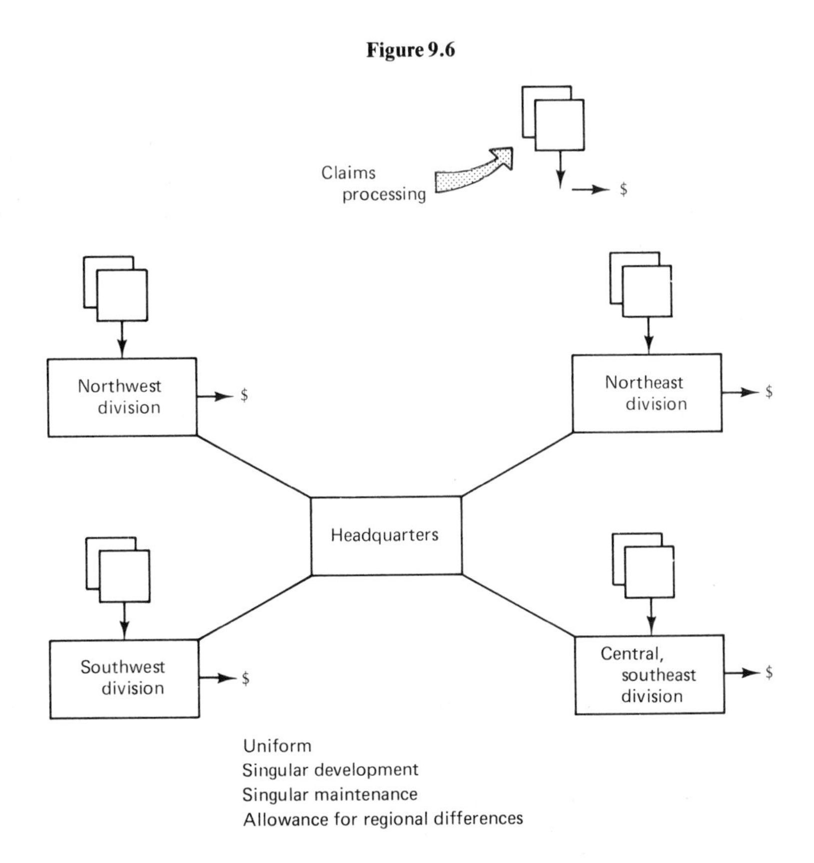

and passes the data to headquarters. As the system is designed, initially no consideration was given to geographic locality and the way in which a claim was settled. At the conceptual level, there is no difference between a claim in the Northeast and a claim in the Southwest. Any function that can be performed in one place can be performed anywhere else. Thus the system is nonredundant at the logical level. One set of programs is developed and maintained. The same source code is passed to each geographic location and is executed there. So at the *physical level* there is a great deal of redundancy.

What about the issue of redundancy between a division and headquarters? If headquarters is able to perform the same function as can be performed in the local areas, the same source code is used in both places. More likely, the activities at headquarters are functionally different from the activities that occur at the division level. When there is a difference in function, there is no redundancy. The usual case is for headquarters to process data in a different *mode* than that used in a division. Typically, headquarters uses data for decision-support systems (to manage the company), while the division uses the same data for operational systems (to run the company). When data and processes are separated by a difference in *mode* of operation, it is normal for redundancy to occur and it is normal for data to flow between the different modes (whereas it is not normal for data to flow otherwise).

As an illustration of the dynamics of managing an environment where data and processes exist in a physically redundant state and in a logically unified state, consider the following few figures. (*Note:* The example used here discusses data. A parallel example could be constructed for the physical form of the processes as well.) Figure 9.7 represents a simple claims system for three divisions of an insurance company. The structure of data is defined identically for all three divisions (and in that sense is redundant) but in terms of content is nonredundant. If a policy exists in division A, it cannot exist in division B and C, for example.

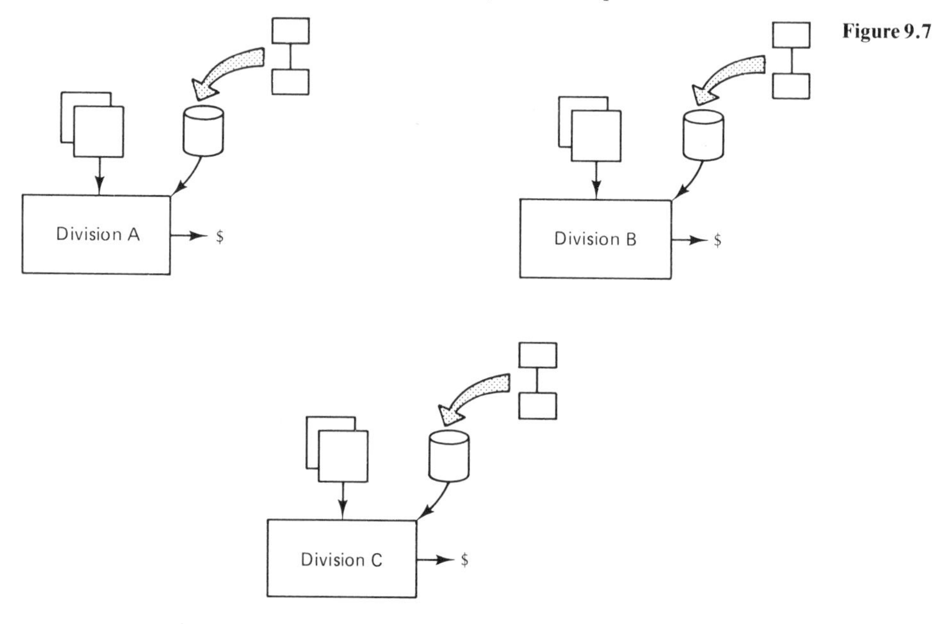

Figure 9.7

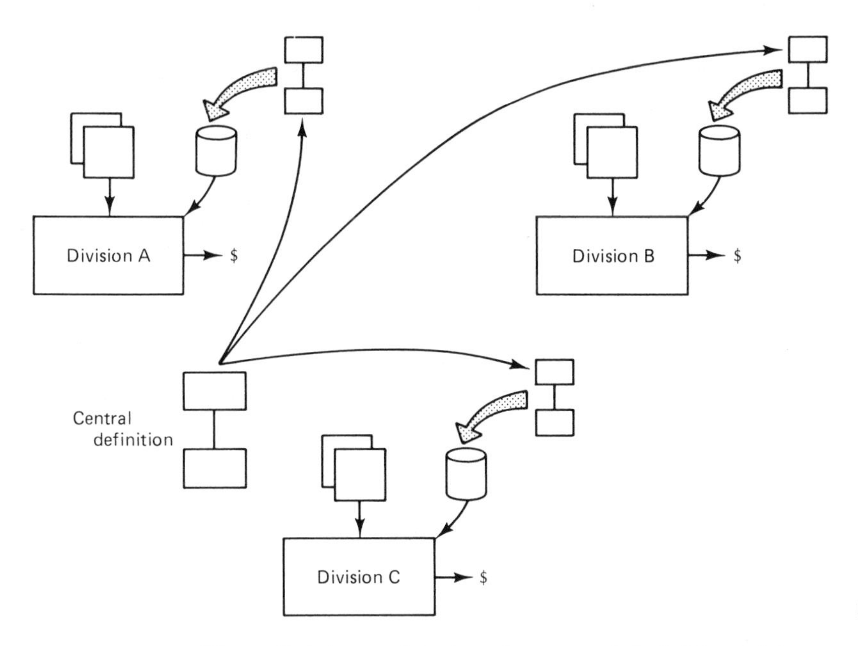

Figure 9.8

Figure 9.8 shows that there is a central definition of data at the source level. The form of the data exists in each division in *exactly* the same form as it exists in the central definition. One day a change must be made to the system. The structure of the data must be changed. Change occurs first to the central definition of the data (its source) as shown by Fig. 9.9. After the central definition is changed, the change is passed to each of the divisions, as shown by Fig. 9.10.

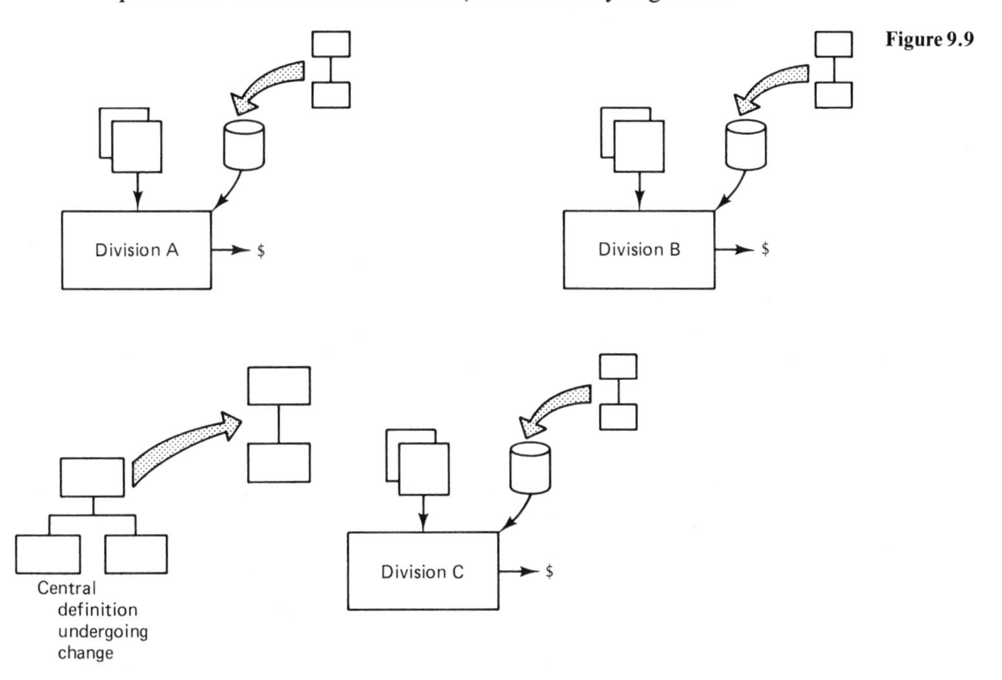

Figure 9.9

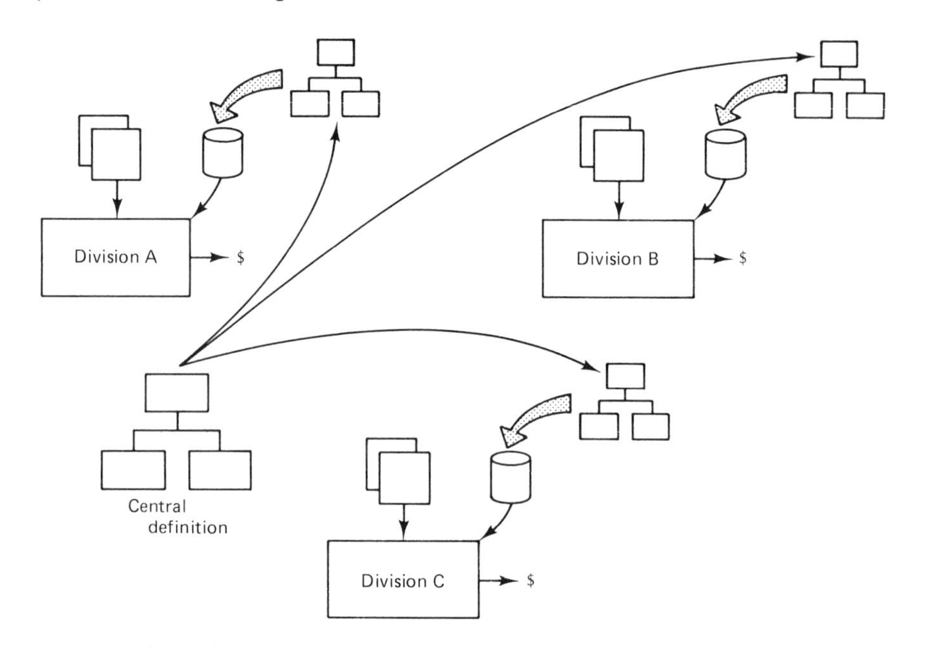

Figure 9.10 Changes to the central definition are reflected at each of the divisions.

Thus it is that development, maintenance, and control are done in a single place (centrally), and the physical structure of the data exists redundantly in multiple places. Exactly the same mode of development, maintenance, and control exist for processes as exists for data.

MULTIPLE PROCESS VIEWS

While the primary business-based view of a process is the foundation for much of the analysis and design that occurs, it is not the only legitimate view of the processes of a company. Consider the two views (i.e., two cycles of business) of an insurance company, an actuarial view and a primary business-based view, depicted in Fig. 9.11. There are many differences between the two views. Which view is correct and how should differences be resolved? *Both* views are correct. Neither view negates the other.

Resolution comes in the consolidation of requirements so that *both* views have their needs satisfied. For example, one of the major activities of the primary business-based system is claims processing. If the following activities were included in claims handling:

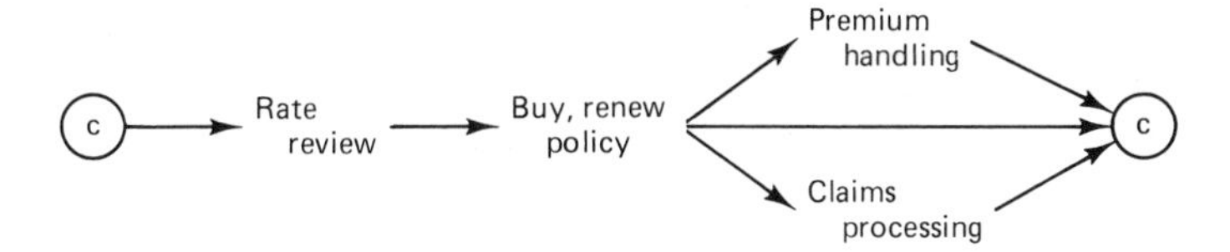

Primary business based view of an insurance company

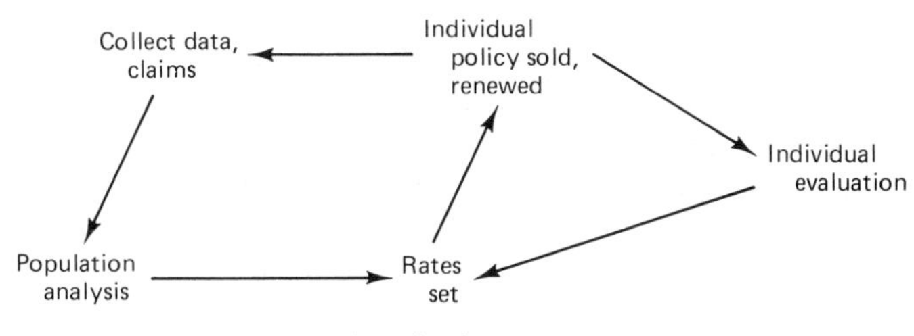

Actuarial view of an insurance company

Figure 9.11

Collect Statistics on:

- Claim numbers
- Size of claim
- Type of claim
- Date of claim
- Policy limitations
- Specific data—age, sex, medical, etc.

then the actuarial *and* primary business-based processing could be satisfied (as far as claims processing is concerned).

The complete major views of processes and data of a company comprise their systems. The most important view is the primary business-based view, but *all* views are necessary to complete the corporate picture.

Once the DPMs have been created for all dimensions, it is normal to have some degree of overlap. To create an integrated process model, the redundancy of the processes must be identified and consolidated. This is done through the consolidated process model. The CPM is nothing more than the collective processing requirements of DPMs that accomplish like functions. The consolidation process amounts to nothing more than an identification of like processes and requirements, including prerequisite and transformation data.

Note that not all DPMs are consolidated but only those sharing a common function. For example, suppose that a DPM exists for processing a loan balance, a savings deposit, and payments to a bankcard. These processes share a common function—the crediting of payments to an account. The consolidated DPM accounts for the difference in type of account serviced while consolidating common activities.

FROM THE LOGICAL TO THE PHYSICAL

The DPM provides a blueprint for the passage from the logical (or conceptual) model of a system to the physical model. There are *many* physical considerations that must be made, most of which are specific to hardware, software, system size, and so on. Because of the enormity of the decisions and their level of detail, they will not be discussed in this book. Instead, some very basic (and very important considerations) will be discussed. The reader is alerted to the fact that the physical building and implementation of processes is its own large subject.

The underlying philosophy in translating the logical model to the physical model is that there is a single unique piece of source code corresponding to a single unique function. As has been discussed, the single piece of source code can be physically implemented in many physical environments.

The first step in the translation of the DPM into a physical model is to divide the system into various major modes of operation. Typically, these modes of operation are:

- Archival
- Operational
- Decision support
- Administrative

The next step is to divide the different modes into a further breakdown, such as:

Operational

- Batch
- On-line
- Mixed mode

The importance of this grouping of modes of operation is illustrated by the example in Fig. 9.12. In this figure functions A through G are designed. For the batch mode the functions are physically combined, in order to minimize the number of times a tape is passed. In the on-line mode, the functions are physically separated in order to minimize the amount of time any one function uses the data base. Note that the *same* function (i.e., how the programs are designed) is accomplished in both cases, but the physical arrangement of the function is quite different.

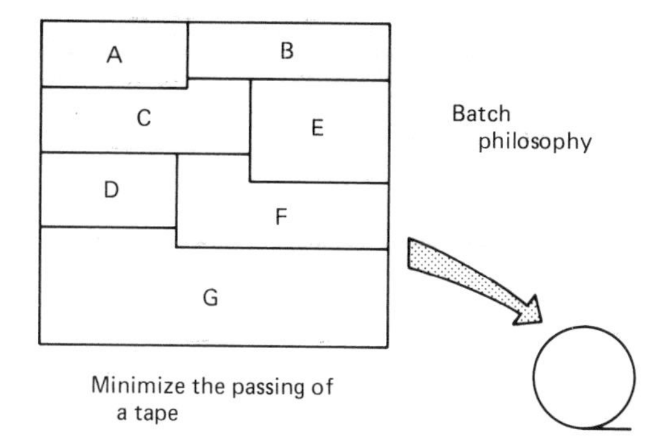

Realized by:
Grouping as many functions together as possible
Once data are arrived at, do as much work as possible before giving up control of the data

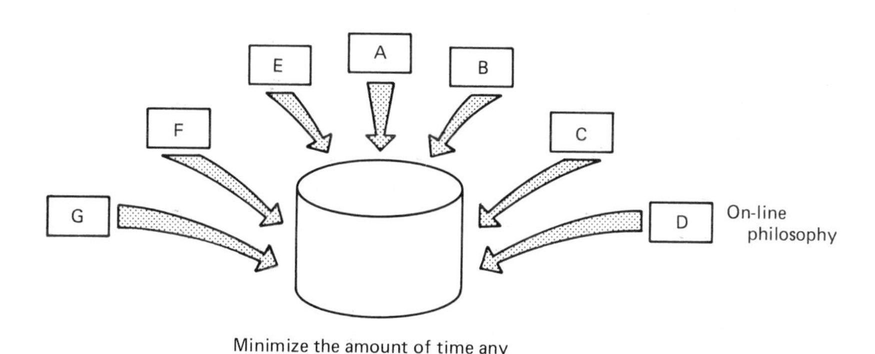

Realized by:
Separating functions into as small and distinct groups as possible

Figure 9.12

BOTTOM-UP PROCESS MODELING

To this point the discussion of process modeling has been from the top down. There is a case to be made for bottom-up process modeling, however. As in the case of data modeling, bottom-up process modeling is both discretionary and useful in gathering details that might have been missed in building the top-down process model. A complete bottom-up process model is usually very cumbersome, very redundant, very complex, and usually reflects the world as it is today, not as it will be in light of a new system.

The steps of building a bottom-up process model are very similar to the steps taken in building a top-down model. The primary difference is in the scope of the model. The scope of integration is used for the top-down model, whereas the scope of a single user's view of the system suffices for an individual bottom-up model. Because of the difference in scope, the bottom-up model is smaller and much less complex.

The selection of which user's view should be modeled bottom up must be done judiciously. Typical selection criteria are (as in the case of bottom-up data modeling):

- Criticality of the user's view
- How representative of the user's view

Once the PVs, DPMs, and consolidated DPMs are merged, the result is the program specification model. The program specification model is the basis for the building of program specifications.

DPM AND PERFORMANCE

Just as performance is an issue at the point of physical data modeling, so performance is an issue in the translation of the process model into program specifications. In the batch environment performance is not a burning issue. But in the on-line environment one of the keys to long-term on-line performance is the design of on-line transactions.

The program specification model outlines the function that must be achieved by the system, but it does not necessarily say anything about the way the function is to be executed. In the on-line environment it is *mandatory* that the function be executed a small set at a time, in order to conform to the standard work unit. For example, suppose that it is desired to search sequentially all the checks that are written by an account for a given month. As long as an account has a normal (30 to 40) number of checks per month, the function can be implemented by simply finding the account, then looking at all checks.

But suppose that a commercial account with 50,000 checks needs to be scanned. Now the packaging of the function is more complex. The performance of the on-line system will be hurt if the function of getting an account, then getting all checks is executed, as many resources will be required to scan 50,000 checks. So the function is packaged differently. A program is written to look at an account and the first 100 checks. Then control is passed to the operator to determine if the next 100 checks need to be scanned. If so, the account is located and processing begins at check 101 and goes to check 200. The first 100 checks which have already been scanned are not searched again. The process iterates until the operator finds the check being looked for or all checks are exhausted.

Note that the business function did not change—an operator can get to an account and then can scan the checks belonging to the account. What changed was the implementation of the function—the work that needed to be done was broken into small pieces.

The packaging of processes in small pieces based on the program specification model is one of the primary keys to on-line performance. As the detailed process model is translated into program specifications, a major concern is the satisfaction of on-line performance.

IN SUMMARY

The detailed process model (DPM) is derived from the functional model in much the same way that the functional model is derived from the BPM. The DPM represents a further breakdown of a process, to a lower level of detail, from which individual program specifications can be made. After the breakdown is complete, data is attached to the process by the collection of data in which the data reside and by data item. The strategy is to build logically nonredundant processes with as much implementation redundancy as is required. This strategy allows for centralized development, maintenance, and control. There are two generic types of redundant processes: utilitarian processes and large major processes that occur a multiple of times.

As in the case of data modeling, the bottom-up approach is used for completeness of detail. It should be done strategically, not for every process.

PROJECT STUDY

The IA team of ABC has finished describing the FM activities and moves directly into DPM modeling activities. At this point the primary business-based FMs are described and are ready for further description. As an example of the DPM exercises, the IA team elects to break up the activities transacted against risk (one of the basic process model processes) into two types: payment and credits. There then is a loan payment, loan credit, bankcard payment, bankcard credit, DDA payment, DDA credit, savings payment, and savings credit. The payment activities are outlined as shown in Fig. 9.13. The credit activities are delineated as shown by Fig. 9.14.

Based on the activities that make up the DPM, common processes are outlined and grouped together, creating the consolidated process model. For example, the credits for the different types of accounts shown in Fig. 9.14 are classified into common functions as shown in Fig. 9.15. These common processes are modeled so that they can be packaged into subroutines. In this figure some functions are common to all credit activities and some functions are common to either loans and bankcards or DDA and savings. The next step in the modeling process is to define at a more detailed level exactly what is accomplished by the function. The outline of the processes are shown in Figs. 9.16, 9.17, and 9.18.

Activity transaction against risk

Loan (payment):

- Determine payment form (check, transfer, cash, etc.)
- Review credentials, ID
- Pay loan
- Subtract payment amount from loan funds
- Record payment

Bankcard (payment):

- Verify amount, date
- Verify limit upon verification
- Verify merchant account
- Deduct merchant's handling charge
- Credit charge fund
- Credit merchant account
- Subtract charge from bankcard fund
- Add charge to cardholder account
- Check card-holder limits
- Record payment

DDA (payment):

- Verify amount, date
- Verify payto account (transfer, direct, clearinghouse, etc.)
- Verify account amount
  - insufficient funds procedure
  - sufficient funds, continue
- Subtract amount from account balance
- Credit payee account
- Calculate account activity charges
- Record payment

Savings (payment):

- Verify amount, date
- Verify savings balance
- Insufficient funds for withdrawal?
- Form of payment desired? (cash, check, transfer, etc.)
- Subtract amount from savings account
- Credit transferred account, if transfer
- Record payment

Figure 9.13

Figure 9.14

Activity transaction against risk

Loan (credit):

- Verify amount, date
- Verify form of payment (cash, check, transfer, etc.)
- Retrieve payment amount, interest
- Retrieve payment due date
- Overdue? – calculate late charges
- Subtract interest, late charges from payment amount
- Subtract principal payment from balance
- Record payment

Bankcard (credit):

- Verify amount, date
- Verify form of payment (cash, check, transfer, etc.)
- Retrieve payment amount, interest, service charges
- Retrieve payment due date
- Overdue? – calculate late charges
- Subtract interest, late charges from payment amount
- Subtract principal payment from balance
- Record payment

DDA (credit):

- Verify amount, date
- Verify form of payment (cash, check, transfer, etc.)
- Retrieve balance
- Credit payment to account balance
- Record payment

Savings (credit):

- Verify amount, date
- Verify form of payment (cash, check, transfer, etc.)
- Retrieve balance
- Credit payment to account balance
- Record payment

The different processes in crediting a loan, bankcard, DDA, or savings account.

Processes that are common across crediting activity:

(all) – Verify amount, date
(all) – Verify form of payment
(loans, bankcard) – Retrieve payment amount, interest
(loans, bankcard) – Calculate charges, interest
(loans, bankcard) – Subtract interest, charges from payment yielding principal payment
Subtract principal payment from balance
(all) – Record payment
(DDA, savings) – Retrieve balance
Credit payment to account balance

Figure 9.15

Verify amount, date common process

Prerequisite data:	Transformation data:
Account	Valid account (y/n)
Amount	Valid amount (y/n)
Date	Valid date (y/n)
Ttype – transaction type – (loan, DDA, savings, bankcard)	Valid transfer (y/n)*
Transfer to (*)	Valid transaction (y/n)
Special handling authorization	

*Optional, depending on activity

Processing:

Valid account, valid amount, valid date, valid transfer, valid transaction = n
Retrieve account
Account valid and existing? if so – account valid = y
Is date valid (yr eq 83, 84, or 85, mo eq 1 through 12, day eq 1 through 31)?
 if so – valid date = y
Is amount valid (amount gt or eq 0, or gt 10,000 and special handling ne 'b̸b̸')?
 if so – valid amount = y
If transfer to ne 'b̸b̸b̸b̸b̸b̸b̸b̸b̸b̸b̸b̸b̸b̸b̸b̸b̸b̸b̸b̸' then – retrieve transfer to account
 if transfer to account is valid and existing then – valid transfer = y
If valid account = y and if valid amount = y and if valid date = y then – valid
 transaction = y
If transfer to ne 'b̸b̸b̸b̸b̸b̸b̸b̸b̸b̸b̸b̸b̸b̸b̸b̸b̸b̸b̸b̸' then if valid transfer = n
 then – valid transaction = n

Figure 9.16

Figure 9.17

Retrieve payment amount, etc. common process

Prerequisite data:	Transformation data: (from null to valued)
Account	Payment amount
Ttype – transaction type – (loan, DDA, savings, bankcard)	Interest
	Late charges
	Service charges

Processing:

Using account as key
 Retrieve payment amount
 Interest
 Late charges
If ttype = 'bc' or if type = 'dd'
 Retrieve service charge

Calculate charges, interest, etc. common processing

Prerequisite data:

Ttype – transaction type –
(loan, DDA, savings, bankcard)
Account
Number of charges
Payment due date
Payment date
Late charge
Transaction payment
Daily rate table
Charge activity amount (n)
Charge activity date (n)
Previous balance
Previous balance date

Transformation data:

Payment due date
Previous balance
Previous balance date
Account balance
Payoff (y/n)

```
Processing:
transaction late charge = 0
payoff = n
if payment date gt payment due date then –
   transaction late charge = late charge
(interest calculation):
transaction interest = 0
if ttype = 'bc' then –
   translate interest rate into daily rate using banks standard tables
   calculate days from payment due date to payment date
   set charge amount from previous amount
   set charge date from previous balance date
   for charge date lt previous balance date –
      i = 1
      for i lt number of charges
         if charge activity date (i) = charge date
            add charge activity amount (i) to charge amount
         i = i + 1
         transaction interest = (charge amount * daily rate) + transaction interest
   add 1 to charge date
   if day gt days in month then –
      add 1 to month
      set day = 1
   if month gt 12 then –
      add 1 to year
      set month to 1
set previous balance date to payment date
if ttype = 'lo' then –
   translate interest into monthly rates using books standard tables
   transaction interest = previous balance * monthly rate
transaction charge = 0
if ttype = 'bc' then --
   transaction charge = number of charges * charge rate
principal payment = transaction payment – (transaction interest + transaction charge + transaction late charge)
account balance = account balance – principal payment
if account balance it or eq 0 then set payoff = y
```

Figure 9.18

Figure 9.16 shows what must be done to verify amount and date for all activities. Figure 9.17 describes what must be done to retrieve amount, interest, late charges, and service charges. Figure 9.18 shows what must be done to calculate certain basic costs, such as interest on bankcard, late charges, and service charges, if any. The form of the DPM is in pseudocode, similar to FORTRAN or PL/I. The prerequisite data and transformation data are shown. It is assumed that prerequisite data are available prior to the execution of the process. Although many programming details are missing, the flow of the algorithm is easily ascertained. Note that the processing specified is simple and limited in scope. Also note that the processing is separated along functional lines (e.g., the verification of amounts and account is separated from the calculation of interest).

While the DPM can be used as a model for program specification (and in fact should be) there are still plenty of programming details that are yet to be specified. Such details occur for loans for a car, a home, or signature loans. For savings details yet to be specified there are passbook savings, money-market savings, and certificate-of-deposit savings. Before program specifications can be written, these levels of detail must be accounted for.

The overall flow of processing for account receiving is shown in Fig. 9.19. In this figure the different processes are defined as they are to be constructed in the primary processing code. The prerequisite data and transformation data are listed, together with the general flow. Using this pattern of different levels of flow, the DPM for processing of all accounts is constructed. Note that the interest calculation code (or any other type of code) exists nonredundantly. If there is a need to reuse the interest calculation code elsewhere, the source can be lifted as a subroutine. Indeed, there should be no bankcard interest calculation elsewhere in ABC.

After the DPM is created from the top-down perspective, an effort is made to verify that all details have been included. It is given that at this point some details

Figure 9.19

Account crediting cycle using common code

```
Processing:
VERIFY AMOUNT, DATE (Prerequisite: . . . ., Transformation: . . . .)
if valid transaction = n then invoke INVALID TRANSACTION PROCESS (Prerequisite: . . . .)
VERIFY FORM OF PAYMENT (Prerequisite: . . . ., Transformation: . . . .)
if ttype = 'bc' or 'lo' then –
   RETRIEVE PAYMENT AMOUNT (Prerequisite: . . . ., Transformation: . . . .)
   CALCULATE BALANCE (Prerequisite: . . . ., Transformation: . . . .)
   if payoff = y then invoke ACCOUNT TERMINATION PROCESS (Prerequisite: . . . .)
if ttype = 'dd' or 'sa' then –
   RETRIEVE BALANCE (Prerequisite: . . . ., Transformation: . . . .)
   CREDIT PAYMENT (Prerequisite: . . . ., Transformation: . . . .)
RECORD PAYMENT
```

Note: Commonly invoked processes are shown in capitals.

(namely those that relate to lower classifications of processes such as car loans, home loans, etc.) are not completely accounted for. Process views (PVs) are created strategically for the purpose of bottom-up verification of the top-down design. The first step in PV creation is to decide which PVs will be created. As in the case of DVs, there are so many PVs that could be modeled and such a degree of overlap that strategic modeling is called for. As an example of strategic creation of PVs the vice-presidents of home loans, car loans, and other retail loans are asked to attend or send a representative to a PV creation session. The representative is high enough in the organization to be able to speak for all loan functions but not so high that day-to-day details escape the person. In some cases more than one representative is sent to the PV session.

At the PV creation session a PV is constructed, much in the format of a BPM. The basic processing cycle is identified, the processes of the cycle are identified, and the flow between the processes is outlined. The modeling process is the same as at higher levels, but the model is created for a lower level and for each different user. In some cases activities such as establishing risk are common and are already covered within the confines of another activity at a higher level. In other cases details of processing are peculiar to a given process view. For example, the collateral and collateral conditions and processing are completely different for car loans, home loans, and signature loans. The PV session makes note of these differences.

After all strategic PVs have been created, the common processing models are enlarged to include the details discovered in building the PVs and the individual differences are documented. From this model program specifications can be written.

1. Construct DPMs for the functions not constructed in detail (i.e., the verification of form of payment, the retrieval of DDA and savings balance, and the recording of payment).
2. What data bases will be needed to satisfy the prerequisite data and transformation data requirements?
3. What data modeling activity corresponds to the DPM activity?
4. Is it possible to create redundant code even when DPMs have been constructed? What can be done?
5. Who should create the DPM? Who should verify its completeness? What guidelines should be used in separating the various functions within the DPM?

EXERCISES

1. (a) What are the differences between the substance of the DPM and the functional model?

(b) What differences exist between the *form* of the DPM and the functional model?

2. Does every function have to be broken down to the DPM level? Why? Why not?

3. Create DPMs for the functional models created in Exercise 3, Chapter 8.
4. Is the efficiency of processing a consideration in organizing the breakdown of processing? When does it become a factor?
5. (a) Discuss several ways that redundant process definitions can be managed at the logical level and at the physical level.
   (b) Why is it important that they be managed at all?
   (c) What happens if they go unmanaged?
6. (a) What are the implications of transporting code across:
      (1) Machines of the same size and make
      (2) Machines of the same size and from different vendors
      (3) Machines of varying sizes
      (4) Machines operating under varying operating systems
   (b) How can physically redundant code be managed in each of these environments?
7. (a) How important is bottom-up modeling?
   (b) When must it be done?
   (c) When should it not be done?
   (d) Who should do it?

10

Marrying the Data and Process Model

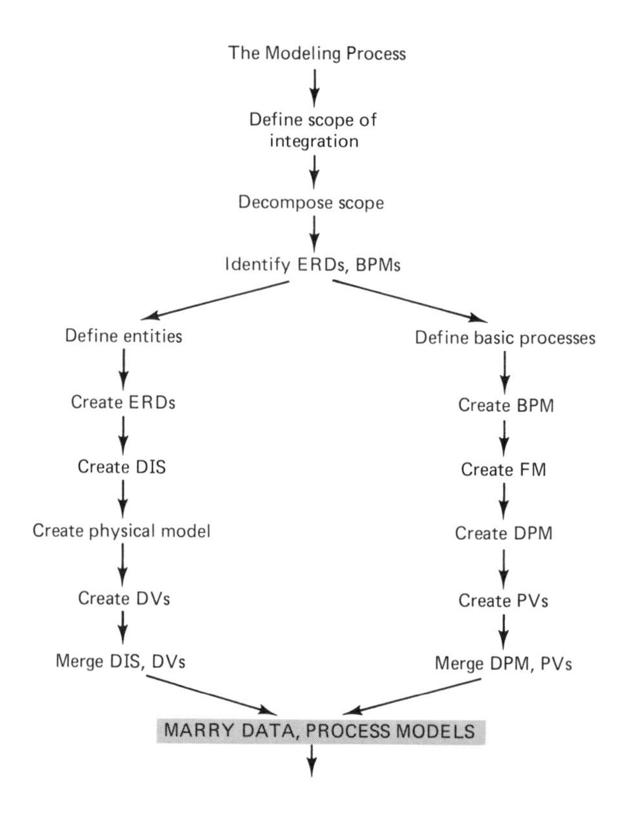

INTERDEPENDENCE OF DATA AND PROCESSES

The relationship between data and process is interrelated and interdependent. A symbolic representation of the interdependence is illustrated by the yin and yang symbols shown in Fig. 10.1. Any design effort that concentrates entirely or mostly on one aspect at the expense of the other risks creating a seriously flawed design. A stable design must represent *both* data and processes.

In the chapter on process design it was seen that after the process was decomposed, analyzed, and organized, reference was made to the data necessary to satisfy the operation of the process. In this case it is clear that there is a close relationship between data and process. However, in the case of data design, the close relationship was not demonstrated, although the relationship in fact exists. Consider the first few steps of data design:

- Scope of integration identification
- Business decomposition
- Entity-relationship derivation

These steps are depicted in Fig. 10.2

In the identification of the scope of integration and the business decomposition, no distinction is made between data and processes. At that level the emphasis is on the system itself—in its entirety. Only at the entity-relationship level is the separation of data and processes made, as the ERD represents data. However, it is implicit that business processes shape the data, whether or not the processes are formally acknowledged. The ERD is a valid description of data only within the context of the processes within which the data is used. For example, consider the simple manufacturing ERD shown in Fig. 10.3. There is no intrinsic relationship between order, part, location, and a recursive relationship among parts (i.e., the data and relationships do not exist in and of themselves). Instead, the data and the relationship are derived from the processes of the manufacturing environment, as shown in Fig. 10.4. Here it is seen that the data and relationships derived from order processing is part and order, the data and relationships from the bill of materials processing is part number and "where used" and "into" relationships, and the data and relationships derived from

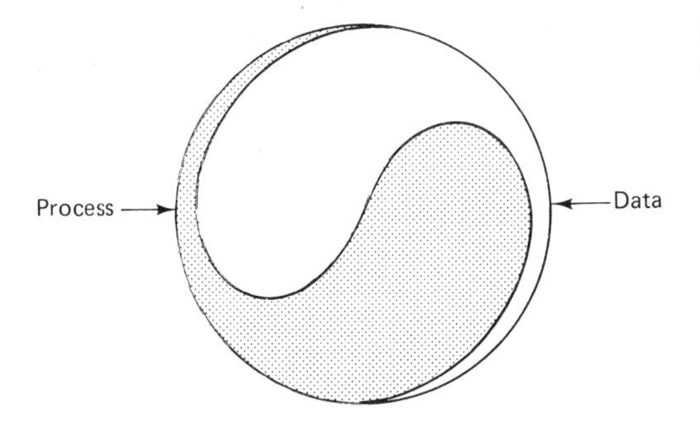

Figure 10.1 The interrelationship of data and processing.

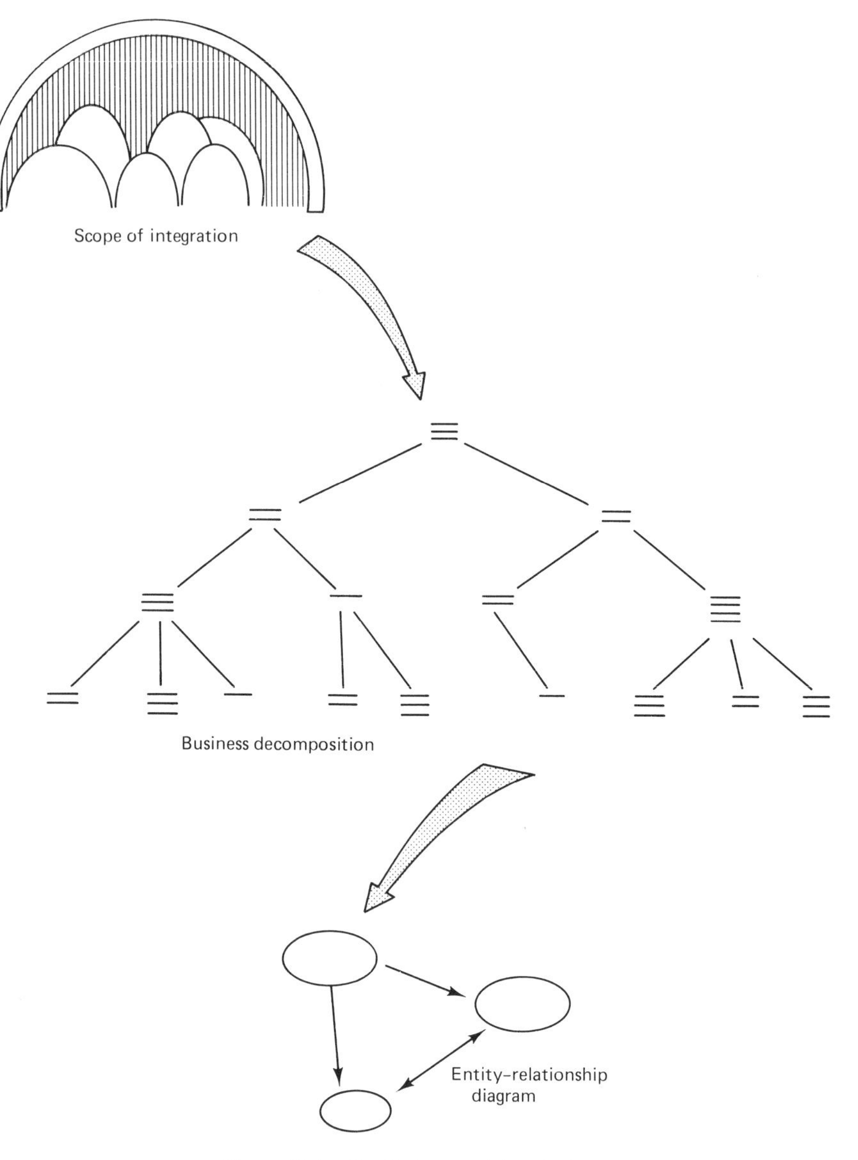

Figure 10.2 The first steps of data design.

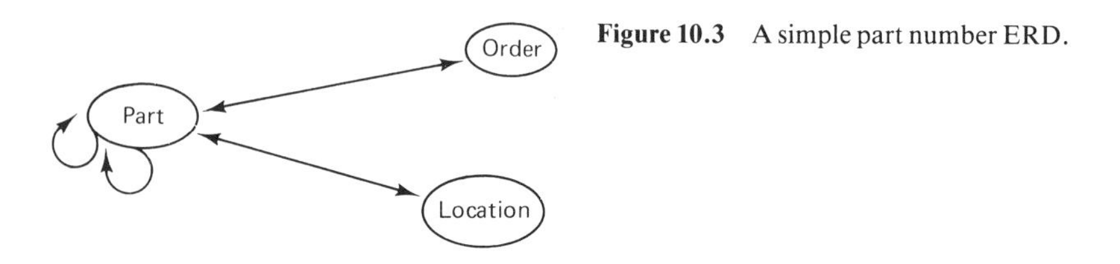

Figure 10.3 A simple part number ERD.

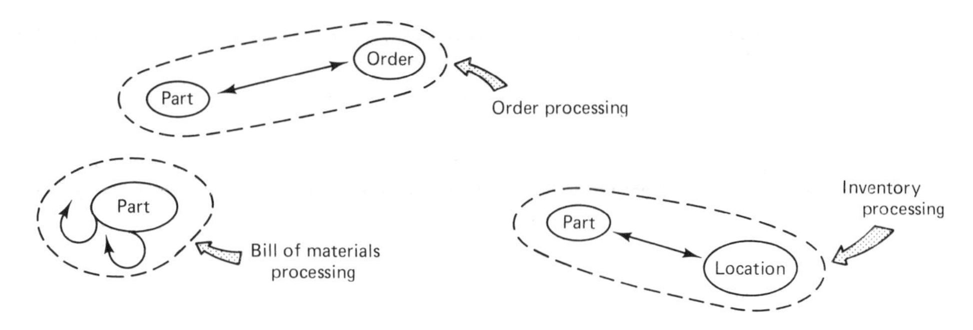

Figure 10.4

inventory processing are part and location. The data do not exist independent of the process, nor do the relationships.

As an example, Fig. 10.5 shows some sample orders, in which several line items exist. The data exist as part of the manufacturer's process and are represented in the ERD as a double-headed arrow, since any order may have multiple parts, and any part may have multiple orders. But suppose that the business of the company was such that only one line per order existed (as is sometimes the case), as shown in Fig. 10.6. In this case the ERD shows a single-headed arrow. Now this is a small, almost trivial point, except that it points up the fact that the *process* (or the business of the company) shapes the ERD. The ERD does not exist in a vacuum.

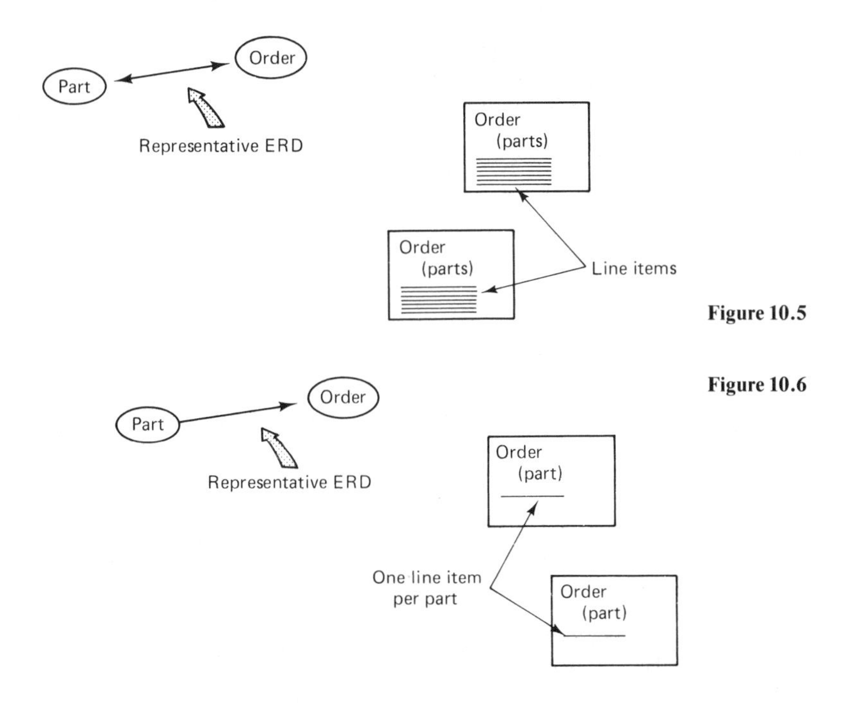

Figure 10.5

Figure 10.6

As another example of an order-processing relationship, consider the order for a ticket to a Dallas Cowboy football game, shown in Fig. 10.7. Here an order is made for a particular game for a seat or a block of seats. There is no location required (since each seat is unique and has a permanent location), and there is no recursive relationship between seats. Also, an order may request multiple seats, but any given seat cannot have two orders for the same date (unless, of course, people want to sit on each other's laps!). The representative ERD is shown, and it is seen to be quite different from the manufacturing ERD, which also processed orders. Again the point is that ERDs do not exist in a vacuum—they directly reflect the business processes they model as well as the data.

The same point can be made about data attributes. No entity "naturally" has a set of data attributes. Like relationships, attributes are dependent on the business process being modeled. To illustrate the dependence of data attributes, consider the order-clearing process modeled in Fig. 10.8, where the only data attributes required for order are amount, authorization, and a supplier relationship. Many other normally appearing attributes for an order (such as line items) do not appear because that is not the business of order clearances. Thus it is that the data attributes depend on the process, even though the business process is not actively illustrated in the data model. In fact, during the data modeling process, the underlying business processes are passively considered, just as data is passively considered in the case of process modeling. The following table illustrates this relationship.

	Data modeling	Process modeling
Active	Data	Process
Passive	Process	Data

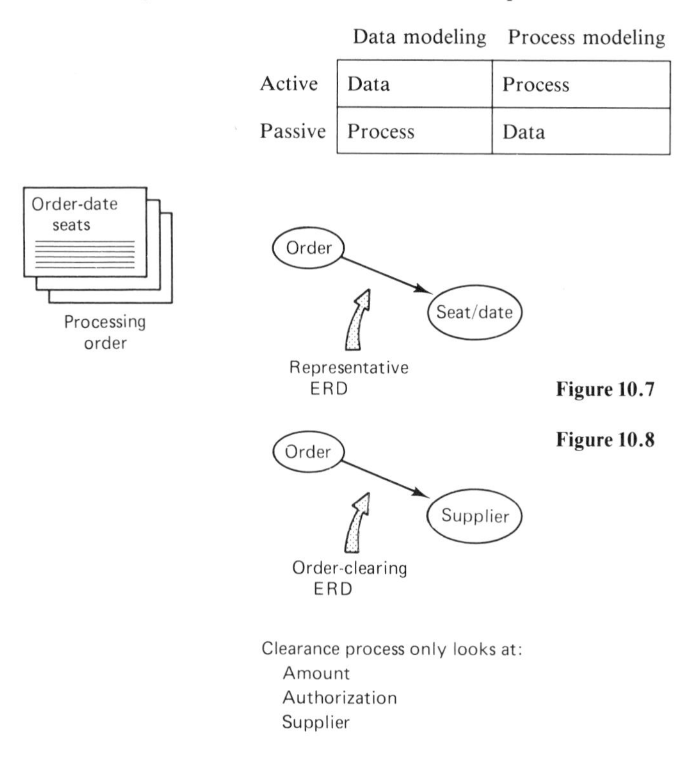

Figure 10.7

Figure 10.8

In fact, there is a corresponding relationship between data and process modeling throughout the modeling of a system. This corresponding relationship is shown in Fig. 10.9. At the highest level of design are ERDs and BPMs. At the middle level of design are DISs and functional models. At the lowest level of design are the physical

Figure 10.9

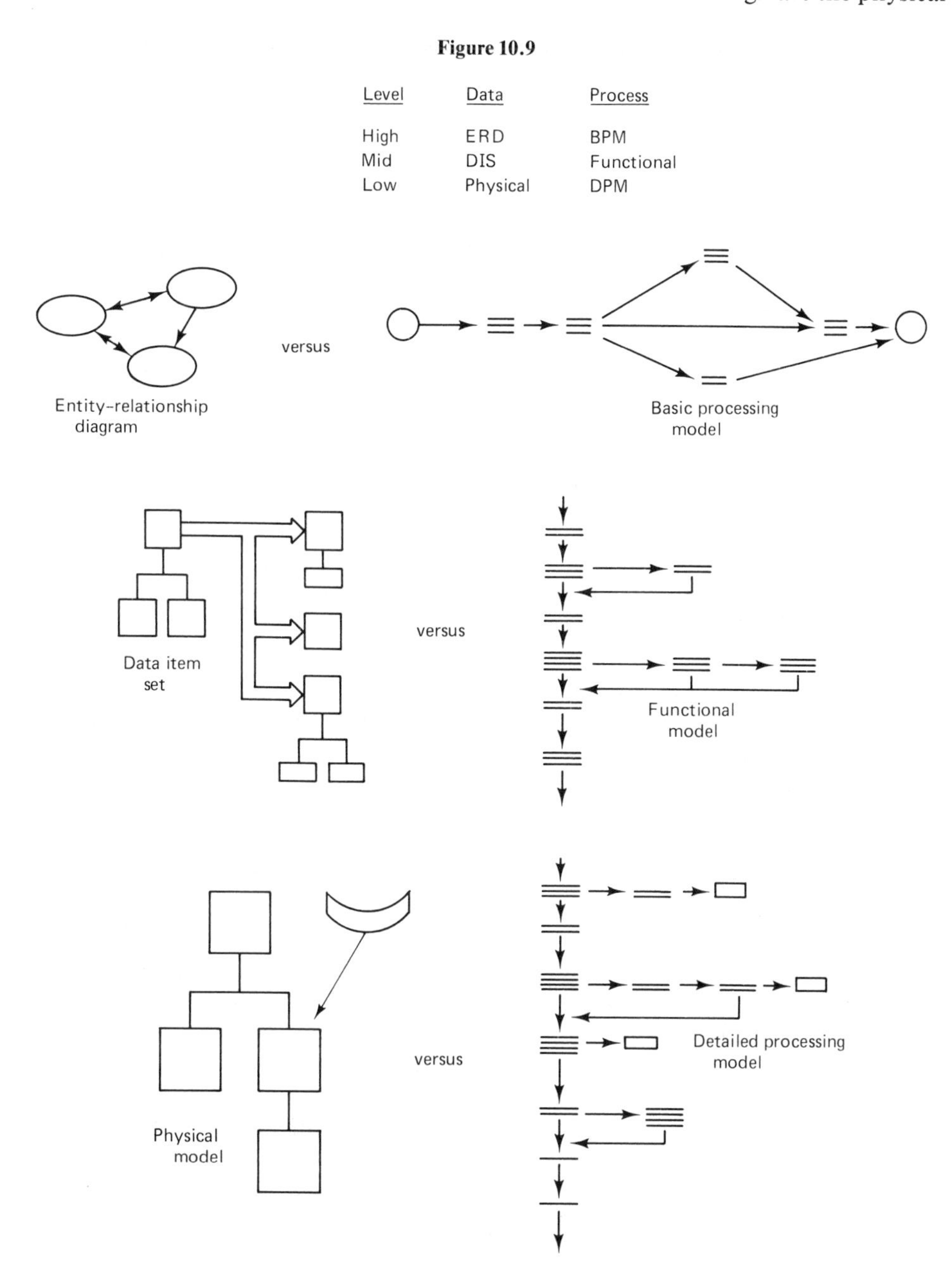

and DPM models. The major activities of design should occur roughly concurrently with each other; that is, ERD construction should be done about the same time as BPM development, for example.

It is clear, then, that there is a constant interrelationship between data and processes throughout the business-based modeling life cycle whether the modeling is done for data or processes. There comes a point, however, where it is necessary to "marry" the data and process models. This usually happens once the physical data model and the DPM have been developed. The "marriage" is merely a process of verifying that:

- All data needed by a process will be available (prerequisite data will be satisfied).
- All processing needed by data will be available.
- No data not needed by a process has been included.

The marriage of data and processes represents a final check on the completeness of the system model. Consequently, the marriage must occur at a low level of detail if it is to be meaningful. The importance of the data and process marriage is underscored by the consequence of what might happen if the check for completeness is not made. On the one hand, it would be possible for data to exist for which there is no supporting process if the marriage validation is not made. On the other hand, a process might exist for which there is no supporting data. Such errors are best caught at the early stages of design rather than in the programming or implementation stage.

What exactly does the marriage process entail? It is simply a matching of the processes against the data, and vice versa. It is best illustrated in terms of an example, as shown in Fig. 10.10. The business that is being modeled in Fig. 10.10 is order fulfillment. The interface with other processes is outlined. The "marriage" process begins where the activities of the order fulfillment are "walked through," making sure that data that need to be present, in fact are present, for the process to function. Each step of the process has its necessary data listed next to its description. Note that *only* order fulfillment is verified; order receipt, customer billing, back-order processing, and production control are not verified. They will be verified individually later. The scope of this verification is strictly order fulfillment. It turns out that all data that are needed is, in fact, available. Had data been needed that was not available, note would have been made and corrective design action would have been taken. If only a few errors were found, the corrective activity would probably consist of the addition of a few data items. Had there been wholesale errors, the DIS or ERD (or BPM or functional modeling) would have to be repeated, properly accounting for the errors that occurred at the marriage.

In Fig. 10.11, data are verified to determine if a process exists to support the data. For the part that order fulfillment plays, the process model is complete. But note that there are many major pieces of data that are not supported by order fulfill-

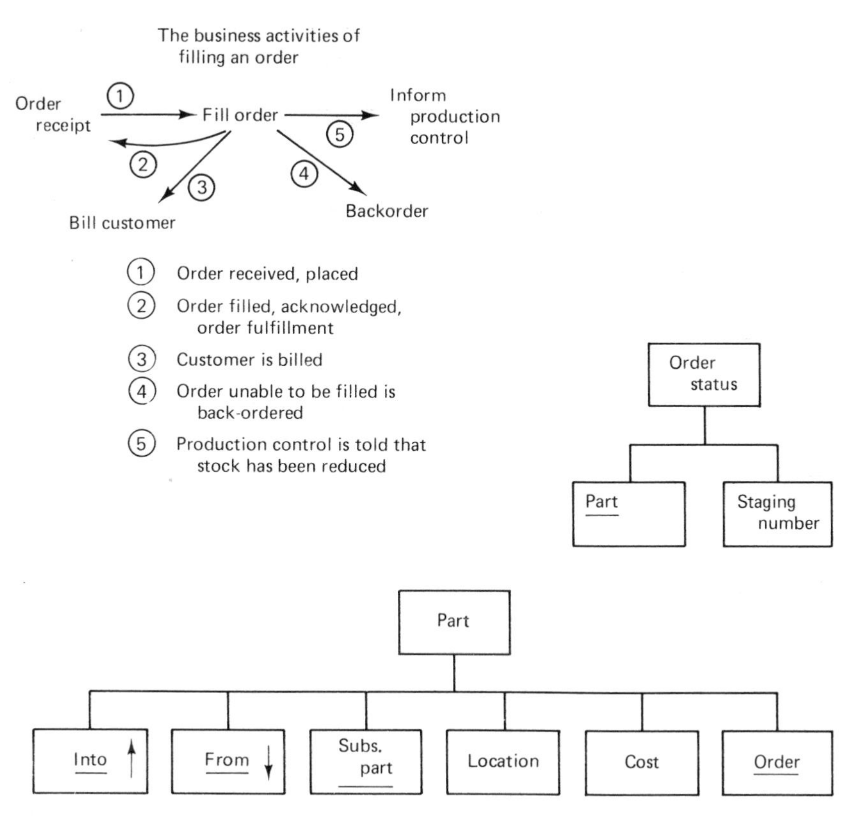

Process-to-data verification

Walking through an order:

Filling process	Data
1. Identify parts (1–n)	Order, part number, amount ordered
2. Create staging number	Staging number, staging data, order
3. For parts 1–n:	Part, location, amount in location
Locate part	Part, amount in location
Allocate part	Stage number, part, amount allocated
Assign part to staging number	Billing flag
Set billing flag	Part, amount, location, date, staging number
Inform production control	
4. If part cannot be located:	Part, location, amount in location
Issue back-order	Date, part, amount, order number (calculated priority)
Assign priority	
5. For billing flags set,	Part, billing flag, amount, staging number, order, date
Issue billing order	Part, billing flag, amount, staging number, order, date

Figure 10.10

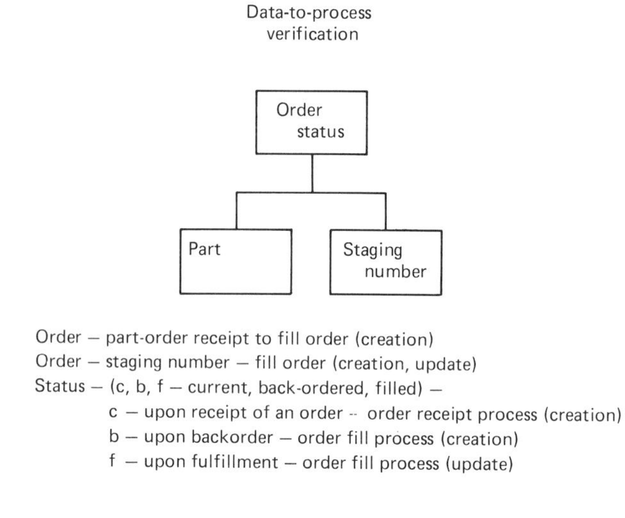

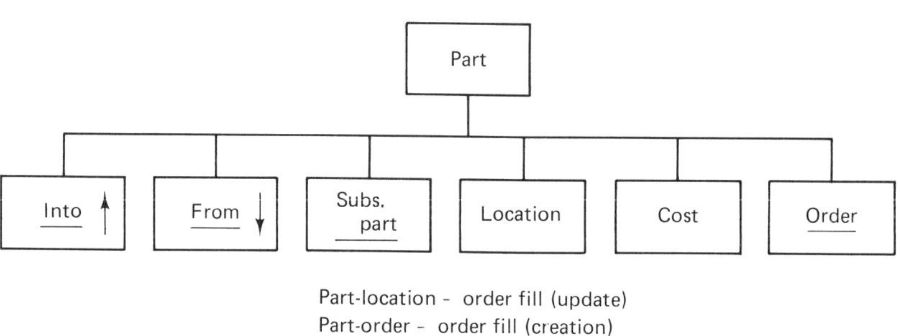

Figure 10.11 Data-to-process verification.

ment, such as INTO, FROM, SUBSTITUTE PART, and so on. These data items are supported by other processing, such as bill of materials. To be complete, *all* data must have its supporting process outlined. Given this example, only order fulfillment is considered.

LIMITATIONS OF ''MARRYING'' DATA AND PROCESSES

Data and process "marriage" is a necessary step in the design of a properly engineered system, but it is not the only step that is necessary (i.e., it is necessary but not sufficient). The marriage process addresses *only* the functionality of the system. It does not address other very important issues, such as system performance or system availability. It is fortunate that these other important issues—performance, and availability—can be addressed primarily at the physical design level which occurs after the marriage. The output of the business-based model can be taken and used as input into the physical design process to produce a system that is:

- Highly flexible
- Business-based
- Performs well
- Available as required

EXERCISES

1. Are there *any* circumstances where data and processes are truly independent of each other?
2. (a) What happens when analysis focuses on data? on processes?
   (b) How can a balance be achieved?
3. (a) Is it possible for data to have intrinsically related attributes?
   (b) What about social security number and name and address?
   (c) What about name, age, and date of birth?
   (d) What about part number and description?
4. (a) Select one of the BPMs and ERDs that have been created from previous exercises and "marry" them. Check for completion and detail.
   (b) Outline where discrepancies have been found and what can be done about it.
5. (a) Who should conduct a walkthrough?
   (b) When should it be conducted?
   (c) Who is responsible for reconciling differences found in the walkthrough?
   (d) What happens if the differences remain unreconciled?

11

Data and Process Modeling and Integration

The preceding chapters on modeling data and on modeling processes have served to describe a unified process for the creation of a stable system model. The chapters have addressed the modeling process at a very basic, step-by-step level in almost a cookbook fashion. The remaining chapters of the book are about the model that has been created, rather than about the methodology that is used to create the model. The remaining chapters draw heavily on the book *Integrating Data Processing Systems: In Theory and in Practice.* Many terms and concepts are referenced from that book. The reader is invited to refer to that book for in-depth discussions.

INTEGRATION WITHOUT MODELS

It is theoretically possible to achieve a high degree of integration without modeling the system formally. However, there are so many obstacles and pitfalls to the achievement of a high degree of integration that it does not make sense to try to integrate without modeling. Following are some of the reasons why system modeling (both data and processes) are essential:

- *Completeness of scope:* Without a top-down system model, the achievement of integration throughout the scope of integration is very difficult. This is so because an integrated environment is a long-term goal that is achieved one step at a time. The system model serves as a road map that guides the development effort over an extended period of time. Without a road map, typically one or two parts of the system are well integrated but other major parts go their own way.

- *Interlocking parts:* Because of the complexity of a large integrated system, many parts of the system (keys, subroutines, shared data, etc.) must be closely intertwined with other parts. Without a model, interlocking the different system parts is difficult.
- *Future additions:* The nature of integrated systems is that they provide a foundation for future additions. Without a model, how and where additions are to be made is open to question.
- *Redundancy recognition:* At the heart of integration is a removal of redundancy of data and processes. Without a model, the commonality of systems (especially where they are large and complex) is not easy to recognize. A system model provides a vehicle for recognition of redundancy.
- *Abstraction:* The analytical tool that is at the heart of integration is that of abstraction. A system model provides a means of abstracting systems.

For these reasons, system modeling serves as a foundation for building an integrated environment. (*Note:* Even in the best of environments, systems modeling is necessary but not sufficient.) System modeling does add more time to the development life cycle, but that amount of time is repaid in multiples in terms of a reduction in unnecessary development and maintenance costs. An unintegrated system, by definition, must be developed piecemeal, with the same function and data being developed multiple times. In addition, maintenance costs multiply as the system ages. But an integrated system is developed and maintained once, thereby costing a fraction of the resources required by the equivalent unintegrated system. Figure 11.1 compares integrated and unintegrated development.

The unit of measure in Fig. 11.1 is in hours consumed in development and maintenance. A set of functions, A, B, C, and D are to be developed. In the integrated case, more work is required initially for total system definition, but much more *total* work is required in the unintegrated case. Of interest is the resources required in maintenance. In the unintegrated case, the maintenance of function A

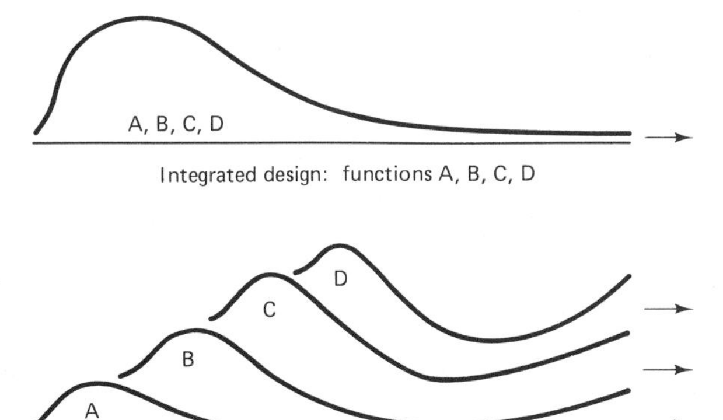

Figure 11.1 Resources consumed: unintegrated versus integrated development.

is fairly normal. But the amount of resources needed for maintenance climbs disproportionately as each new function is implemented. There are several reasons for this phenomenon. The primary reason is that maintenance of a new function must be coordinated with other functions. The duplication of data requires that code be written for the express purpose of keeping data synchronized, a task not required in an integrated environment.

USING SYSTEM MODELS

The system models that are produced address different levels of the system, as depicted analogically in Fig. 11.2. In this figure ERD and BPMs are shown at the highest level, corresponding to a globe. At the next level, DIS and functional models are

Figure 11.2 Levels of system models.

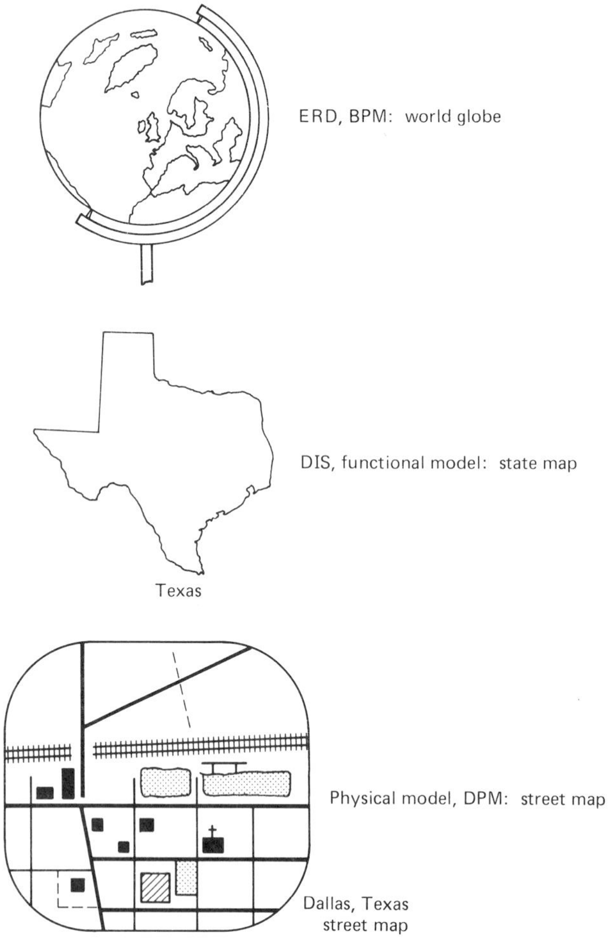

ERD, BPM: world globe

DIS, functional model: state map

Texas

Physical model, DPM: street map

Dallas, Texas
street map

shown corresponding to a state map, and at the lowest level the physical model and DPM correspond to a city street map. The system models then provide a road map to the building of the integrated system. Because of the size and complexity of the integrated environment, the phasing of projects is normal. Consider the problems of development in a phased and a nonphased environment, as shown by Fig. 11.3.

In the phased approach of Fig. 11.3, accounts processing is built first (with savings and money-market requirements in mind), then savings processing is built in conjunction with existing account processing. Finally, money-market processing is built on top of existing savings and account processing. The result is one system that accomplishes all necessary functions.

Now consider the nonphased approach as shown in Fig. 11.3. The first system to be built is the money-market system, with its elements of savings and account processing. Next, the savings system is built, with its elements of account processing. Finally, the account processing system is built. Three separate systems are created and must be maintained. Data and processes must be maintained and synchronized in three places.

It seems obvious that an integrated, phased approach is the best from the perspective of resources—total resources used in development and maintenance. But most systems are not built this way. The reason most systems are not built in an integrated, phased approach is that phasing and integration require:

- *Discipline:* a change in the way an organization builds systems
- *Planning:* an ability to view and act on the system from a global viewpoint

Figure 11.3 Three functions must be integrated: account processing, savings processing, and money-market processing. Money-market processing depends on some aspects of savings, and both money-market and savings depend on some aspects of account processing.

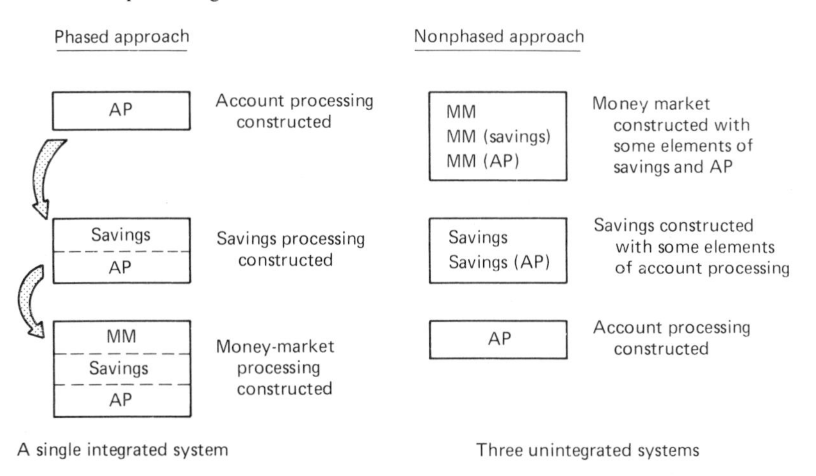

- *Long-term commitment:* an understanding of what is required in the long term and the ability to translate long-term goals into a series of short-term objectives
- *Political unity and coordination:* a desire to do what is best for the shop and not what necessarily satisfies parochial desires

The foregoing realities *usually* overrule the best intentions to achieve a high degree of integration, *despite* the enormous costs of unnecessary development and maintenance.

CONFORMING TO THE GLOBAL MODEL

How exactly does phase-by-phase construction of a system occur, in light of a global model? (*Note:* The following example is done in terms of a data model. The reader is alerted to the act that corresponding activity occurs at the process level as well.)

Using the example of savings, money market, and account processing, a global data model is developed, as seen in Fig. 11.4. This model satisfies *all* data requirements for all three phases. The first phase to be built is the account processings phase. Data is defined as shown in Fig. 11.5. Next, the savings phase is built. This requires using existing space that has been preallocated, and adding a new segment, as seen in Fig. 11.6. Finally, the money-market processing is added. This entails adding more data as a dependent to account, as shown in Fig. 11.7.

Figure 11.4 Phased physical data structure.

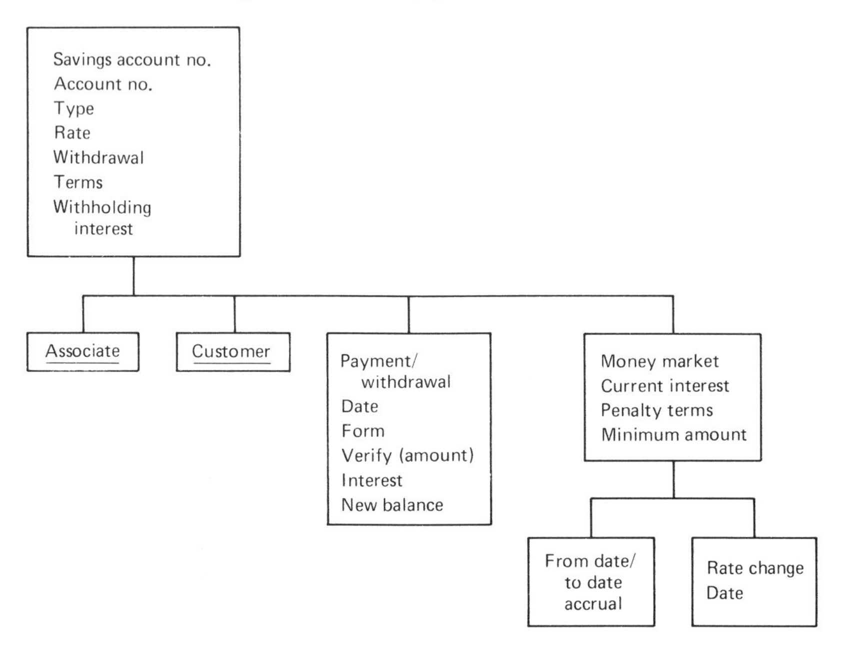

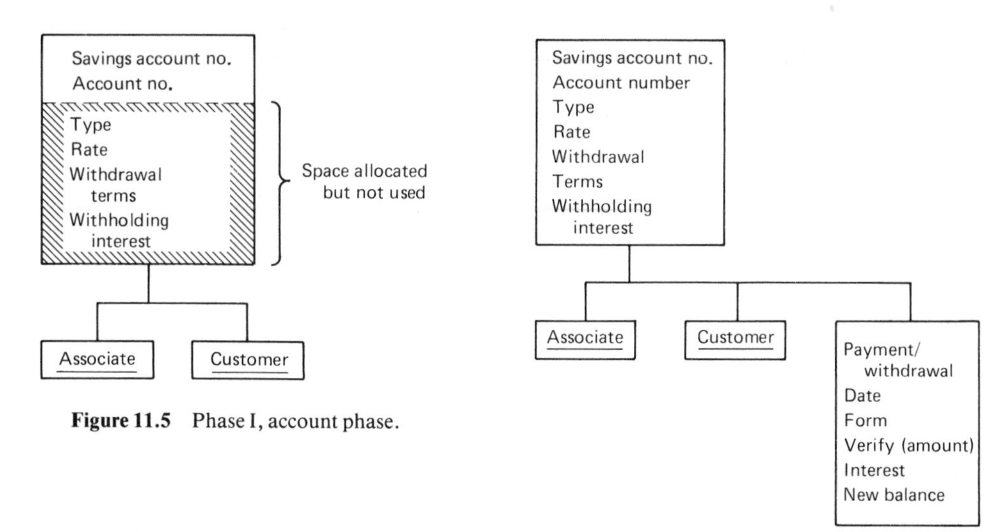

Figure 11.5 Phase I, account phase.

Figure 11.6 Phase II, savings phase.

Figure 11.7 Phase III, money-market phase.

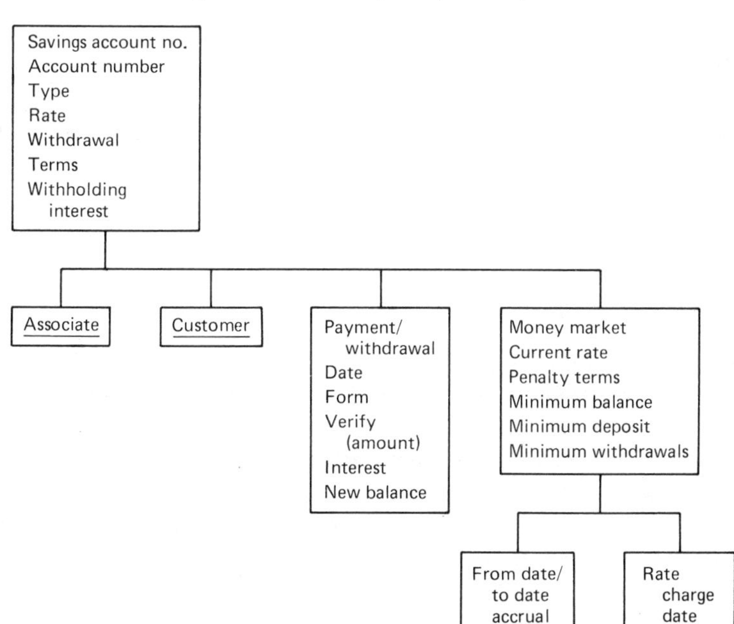

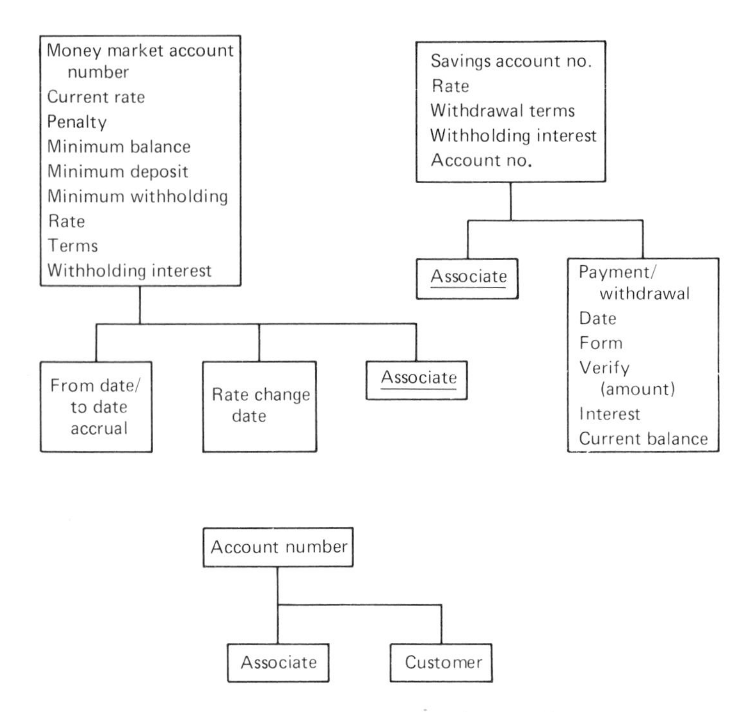

Figure 11.8 Typical nonphased, unintegrated system.

The addition of data, phase by phase, is orderly causing minimum disruption to existing data and code. For an example of how the data might look if it were integrated and not phased, refer to Fig. 11.8. Actually, this figure shows data in its *best* case, where there is a minimum of redundant data definitions and function definitions. Other problems occur in going from account number to savings account number, in the compatibility of account numbers, and in the I/Os done in processing money-market and savings functions.

CHANGING THE SCOPE OF INTEGRATION

Over time, slight adjustments to the scope of integration are normal. Slight adjustments should pose no major problem. What poses problems are large adjustments to the scope of integration. These may occur because of major changes in business or because the scope was improperly defined at the outset. In any case a major shift in scope poses a major problem.

As the scope is adjusted, every attempt must be made to use existing code and data. If a concerted effort is not made, there is little reason to try to integrate. A large shift in the scope of integration may even require the system to be remodeled.

MODELING AND THE SPECTRUM OF INTEGRATION

The end result of modeling data and processes is unification, as described in Section 1 of *Integrating Data Processing Systems: In Theory and in Practice.* The unification of the data and processes leads to a high degree of integration, usually stage 6 or stage 8. The lower stages of integration—1 through 5—are not possible in light of singular process and data definitions. Also of interest is the fact that data can be unified to stage 6 or 8 and be centralized or decentralized. The physical organization of systems (whether they are centralized or decentralized) is independent of the unification of data and processes.

THE CHANDELIER PERSPECTIVE AND MODELING

Integated data is viewed in at least three dimensions, as outlined by the chandelier perspective. This is realized by a data model in that the consolidation of ERDs, DISs, and physical models reflects the different views of data. A properly constructed model accounts for *all* dimensions of data in a consolidated fashion. The many views of the same data that can be captured in a model are illustrated in Fig. 11.9. Each perspective views the same data in a different way. Yet the data remains the same. Such a multidimensional perspective of data is captured by the integrated model. Without an integrated modeling of data, the policy must be viewed separately for the many dimensions in which it exists.

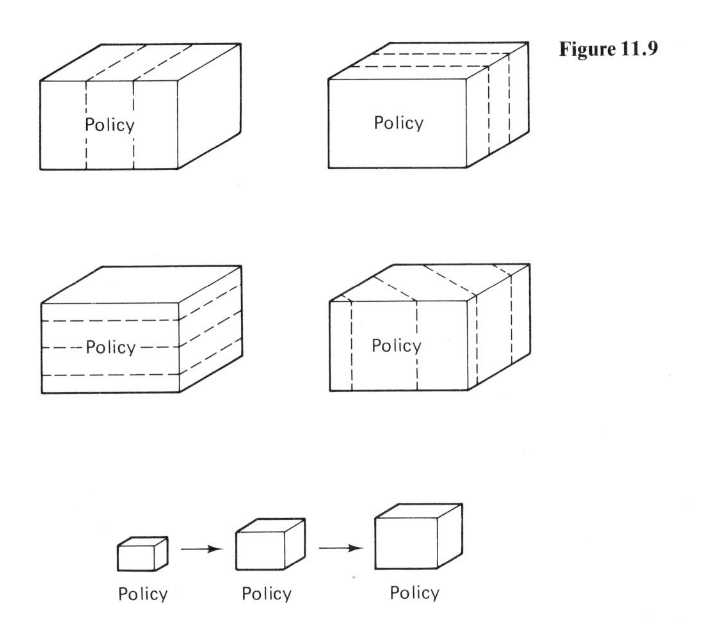

Figure 11.9

USING THE SYSTEM MODEL

Once the system model is created, how can it be used to maintain a high degree of integration? The following steps can be used when it comes time to construct new code:

- Look at the model (of processes) and see if the process exists.
- If the process is not located, contact data administration.
- If the process is located, has it been set into code?
  If yes, is the code usable?
  If yes and the code is not usable, why not?
  If no, set the procedure into code, taking all requirements into account.
- Put code into the public library.

The effect of the foregoing usage of a system model is:

- Code is written once.
- Code is maintained in a single place.
- Code is written for future reuse.
- Code is usable on any processor.
- Code is written in alignment with the long-term plan as reflected by the model.

PITFALLS

Although modeling a system is the key to a methodical discovery of the commonalities of systems, there are several pitfalls to be avoided. Most of the pitfalls relate to the actual execution of the modeling effort. Some pitfalls are:

- Taking too long. A modeling effort that drags on loses momentum, whatever else is done correctly. The modeling effort should be done as crisply as possible.
- Streamlining the modeling process. It is easy to get caught up in detail and in technique. The best modeling techniques are those that are the most direct. Techniques, form, richness of tools, etc. amount to nothing if they do not contribute to the ultimate product—an engineered, integrated system.
- Not choosing the proper personnel for the staffing of the information system architecture project. With the best plan possible, the wrong personnel will short-circuit the effort. Ideally, the information system architecture staff should include both data processing personnel and users, represent people that have the vision of management and the understanding of details of a clerk, are not politically threatening but are politically respected, are able to dedicate time to the effort as needed, and work well together.
- Not setting short-range and intermediate-range milestones as well as long-term milestones.

EXERCISES

1. (a) What are the difficulties in attempting to integrate without a system model?
   (b) What degree of integration is likely to be achieved?
   (c) Can a long-term blueprint be laid without a model?
   (d) Can long-term goals be clearly defined?
2. Lay out a PERT chart for building an integrated system. At the least include ERDs, DISs, physical models, BPMs, functional models, and DPMs. Include other activities, such as functional decompositions, scope of integration definitions, and program specification.
3. (a) Is it mandatory to use the phased approach in building integrated systems?
   (b) What are the issues?
   (c) What is risked?
4. (a) How can the data administration check to see that future designs and future phases will conform to a global model?
   (b) Will control be automatic or administrative?
5. (a) What happens when the scope of integration significantly expands after part or all of the system has been built?
   (b) What happens if the scope of integration contracts?
   (c) What costs are associated with an expansion or contraction?

12

Elasticity and Data and Processing Models

Change in the user's environment is a fact of life. The only difference between users is in the rate of change. Given enough time, every user's environment will undergo change, and that change must be reflected in the systems that support the user. The system designer gives himself or herself the best chance at being able to absorb change gracefully by basing the design of the system on a business-based model of data and processes. The designer stands the least chance of withstanding change by basing the system design on a single set of processing requirements. The difficulties of change that arise when only a single set of processing requirements are considered are:

- *Requirements change:* The impact of requirements change is the heaviest where data is considered to be a by-product of the processing that is to be accomplished (as in the sequential environment).
- *Incompleteness:* A single set of requirements represents only a single view of processing that is to be done.
- *Environmental factors:* When only a single set of requirements is considered, it is difficult to judge accurately how the system should fit with other systems.
- *Future requirements:* Single sets of requirements seldom take into account future processing requirements. The normal case is for only the immediate set of requirements to be considered.

ADDRESSING ELASTICITY

The issue of system elasticity centers around the definition (the source code, the data structures) of the data and processes that make up the system. It does not center around the implementation of the design beyond the point of definition. The degree of elasticity is determined by the design practices that go into *defining* the source code. This means that once the source code and data structures are defined, elasticity cannot be retrofitted into the system without rearchitecting the system at the source level.

Why is the elasticity of a system determined at the point of source definition? There are three principal reasons. When the source code and data structures of a system are built on a business-based model of the system:

- When change occurs in the source code and data structures, they occur in a minimum number of places.
- When change occurs in the source code and data structures, only those elements and code that relate to the changed part of the user's environment are changed.
- When change occurs in the source code and data structures, it occurs in the least vulnerable places.

The minimization of code and data impact by modeling data and processes is achieved because elements and code are not defined redundantly. This limits the proliferation of change throughout the data structure and programs. The process of separation and grouping of data elements and code according to its usage allows change to be localized. The emphasis on top-down design and completeness of scope of integration focuses change onto the least vulnerable part of the source code definition and data structures.

As an example of the differences in elasticity between a modeled and an unmodeled (or integrated and an unintegrated) system, consider the system shown in Figs. 12.1 and 12.2. In Fig. 12.1 data definitions exist redundantly in three physically separate data bases, but there is a single source definition. In Fig. 12.2 data exists redundantly (in definition and content) across six data bases but is *not* controlled at the source level. The maintenance and development (and to some extent, the operation) of the systems shown in the integrated case are significantly less than the resources required for the unintegrated case.

These differences can be viewed at a higher system level as well. Consider three banks A, B, and C that have built systems as shown in Fig. 12.3. This figure shows that bank A has built many systems in a piecemeal fashion over the years. Little thought or commitment has been given to a consolidated plan. Consequently, the individual systems are complex, disorganized, and inconsistent. Bank B has taken a more orderly approach. Loans and savings have been consolidated into single systems. Within loans, for instance, there is uniformity in the way that a customer is treated and there is uniformity in communications between different types of loan

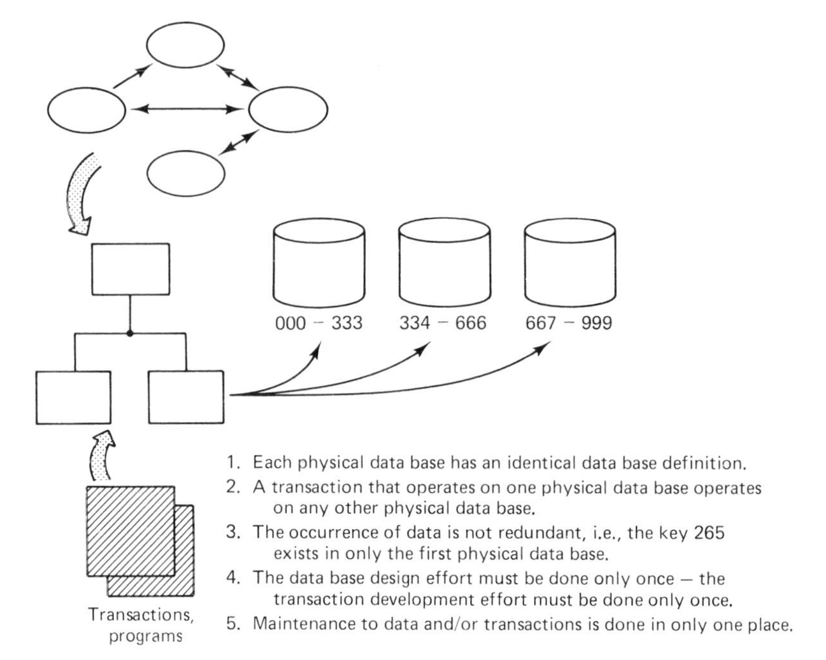

Figure 12.1

Figure 12.2

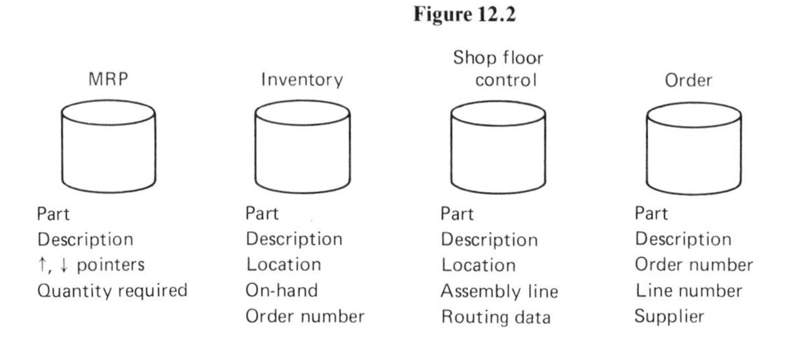

BOM

Part
Description
↑, ↓ pointers
EC notices

1. Redundant data – description of part XYZ exists in five different places.
2. The definition of the data is semantically different in each system (i.e., the data layout is inconsistent).
3. Transactions that operate on one data base will not operate on another.
4. Five development efforts are required; five maintenance efforts are required.
5. When the contents of description for part XYZ are changed, the change must be effected in five places.
6. When the value for part XYZ does not agree in one data base with another data base, the means for resolution are not clear.

Bank A:	Bank B:	Bank C:
Highly unintegrated	Unintegrated	Integrated
Systems	Systems	Systems
Home loan	Loan	Bank customer
Car loan	Savings	Bank services
Commercial loan	DDA	
Wholesale loan	Trust	
Personal loan		
Savings account		
Money-market account		
CD account		
Credit card		
Christmas club account		
DDA		
Personal banking		

Figure 12.3

processing. However, bank B's systems are still not integrated to the highest degree, in that DDA, savings, and loans are separate. Bank C, however, has based its systems on an integrated model of data centering around customer and service, and has achieved a high degree of integration. Note that *functionally,* there is *no difference* in the basic services rendered by A, B, or C. However, the development effort and maintenance posture is *vastly* different at each bank, and the difference is noticeable by the customer when the customer desires to have activities cross functional boundaries.

Consider the impact of a simple change. Bank marketing dictates that customers be grouped into one of three catagories:

- Upwardly mobile
- Preferred
- Regular

This grouping is to be based on such data as date of account opening, average income, average account balance, and so on. To implement this change in the bank's systems, two types of activities must occur:

- Customer data must be gathered and evaluated.
- Rating must be carried with customer data.

In bank A, *each* system must be analyzed and individually changed. Complicating matters is the fact that data must be coordinated across systems (i.e., a DDA customer may be upwardly mobile because of the DDA balance, but may be classified as regular because of loan balances). Each loan system, each savings system, and the DDA system must be analyzed and modified. The amount of work is *huge.*

In bank B, systems are integrated to the functional application level, but there still is no common definition or treatment of customer across loans, savings, and DDA systems. The amount of work to be done is significant.

In bank C, systems are integrated. There is a single customer model with consistent treatment across all functions. The change is easy to analyze and easy to effect. An estimated amount of resources is shown in Fig. 12.4.

For a much more detailed explanation of the differences in development and maintenance resource consumption, refer to Appendix B of *Integrating Data Processing Systems: In Theory and in Practice.*

Why is data and process modeling (and the resulting high degree of integration) so effective in minimizing the resources needed for development and maintenance? There are two primary reasons:

1. Data models, when used as a guide for building systems, reduce the multipliers of development and maintenance—program and data definitions—and localize the impact of change—where there is a single definition of data or processes, there is a single change that is necessary. Where there are multiple definitions, multiple changes are required.
2. Data models greatly reduce the vulnerability of a model to change—the model covers a wide scope, all major keys are identified, and traversals across the systems built based on the model are able to be effected.

One effect of modeling is to group like data and separate unlike data. However, the degree of disjointedness that is achieved is up to the designer. Data modeling does not mechanically replace judgment. In the final analysis, even the most sincerely applied techniques of data modeling are not better than the designer applying those techniques.

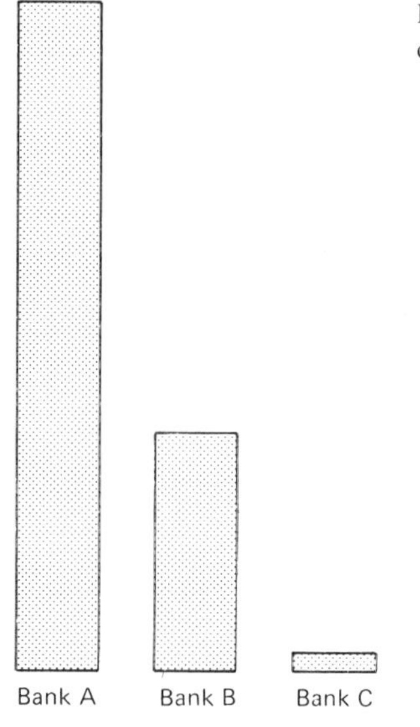

Figure 12.4 Resources required for customer change.

EXERCISES

1. Describe the probability of change in the following environments:
   (a) Bank
   (b) Chocolate chip manufacturer
   (c) Life insurance company
   (d) Auto manufacturer
   (e) Regulated utility
   (f) Government accounting office
   (g) Internal Revenue Service
   (h) U.S. Army
2. (a) What is the impact of change on a system at each of the following levels?
      (1) The program level
      (2) The data definition level
      (3) The conversion program level
      (4) The operational level
      (5) The user level
      (6) The management level

   (b) Where is change most costly?
   (c) What can be done to minimize change?
3. (a) What is system elasticity?
   (b) Why is system elasticity determined at the source level (for code and data)?
   (c) How do the source-level definitions relate to the system model?
   (d) What guidelines should the designer follow to maximize system elasticity at the source level?

13

Politics and Data and Processing Modeling

System modeling is an organizational issue as well as a technical issue. The organizational structure and responsibilities at the user, application, and management levels are normally rearranged or at least fundamentally questioned by system modeling. Despite the fact that system modeling is much to the long-term benefit of the organization, it is usually profoundly resisted at the outset.

The end objective of system modeling is a high degree of system elasticity, which results in a minimum of development and maintenance. The achievement of elasticity comes from a global perspective—one that looks first at the business or organization in its entirety, and second, at the individual components of the organization or business. The traditional local perspective of the data processing organization is not adequate for, nor appropriate to, the construction of integrated systems.

The global perspective must first be recognized and supported at the management level. Management *must* lead the organization into the integrated environment. Why system modeling must be done, what to do to achieve an adequate model, and what to do with the model once built must be subjects with which management is intimately acquainted. It is not enough that management know some vague concepts and buzzwords. The initial organizational resistance is so deep and widespread that management will be *required* to support very actively the effort to integrate.

RESISTANCE TO ORGANIZATIONAL CHANGE

Most organizations resist the effort to integrate because their entire history has been spent in the "local" mode—building systems in a piecemeal fashion for a limited set of requirements. Piecemeal development is very appealing because:

- It requires no long-term plan or commitment.
- It fits development schedules that are driven by individual users and applications development groups acting independently.
- Users and developers feel that they have control over their own data and processes (at the expense of redundancy on a global scale).
- The immediate set of requirements can be addressed without regard for larger system requirements.
- Different users do not have to collaborate and cooperate with other users at the point of system operation.

The result of piecemeal development is a collection of local views of the business of the organization, not a single cohesive view. To achieve integration, users, applications, and management must view themselves and their role in the business as a part of a whole, not their own independent unit. When users, applications, or management put their individual goals and self-importance above that of the organization, the result is unintegration.

Resistance from the user is normal. An integrated effort at the user level usually carries with it a change in responsibilities. These shifts in responsibility are the basis for most of the dissension at the user level. The user's sense of system ownership is threatened by a rearrangement of responsibility that often comes with integration. The fact that the shifts in responsibility are for the long-term good of the organization have little sway with the user.

Resistance from applications stems from the traditional user-applications relationship. Since it is the job of applications to keep the user happy, a perceived threat to the user is indirectly a perceived threat to the applications. In that light, integration requires that application development be done in a manner foreign to traditional development. The changes in the way systems are built requires applications to share design decisions across organizational boundaries, something most applications avoid at all costs when operating in the local mode.

WHY MODEL SYSTEMS?

If systems modeling and the ensuing integration of systems is actively and strongly resisted throughout the organization, is there a real justification for system modeling? The issue of business-based system modeling and integration ultimately boil down to one common denominator—*money*. The *costs* of development and maintenance are such that the effort to integrate is *easily justified*, despite the initial organizational resistance. In an immature organization where workers selfishly cling to traditional, local development practices, the cost of development and maintenance is destined to rise at an ever-increasing rate. In a mature organization where workers focus on the global needs of the organization rather than on limited individual ambitions and are willing to adopt development practices that serve the organization globally, development and maintenance costs can be controlled.

Even in the best-intentioned shop, achieving an integrated environment is difficult because integration is a long-term proposition. The long-term goals of integration must be constantly broken into a series of short-term tasks. Thus it is that implementation of integration is not an easy thing.

REALIGNMENT OF ORGANIZATIONAL RESPONSIBILITY

The first (and probably most profound) change required as an organization heads for an integrated environment is that of realignment of responsibility. The realignment is not nearly as bad as that feared by the user and application once done; the adjustment of attitudes from the local perspective to the global is the most traumatic.

Once domains and responsibilities have been realigned, organization design and development practices must be adjusted. Part of this adjustment is the incorporation of business-based system modeling techniques. Periodic assessment of current activities, past achievements, and future directions are required, especially in light of the long-term nature of integration.

To make these changes in attitudes, practices, and techniques requires organizational discipline. On those occasions where data processing management *must* take a firm stand, management must stand with resolve.

Organizational Assignments

Even though there is no set of hard-and-fast rules as to who does what, the following list represents a general assignment of duties:

- *Data modeling:* data administration, application analyst, user liaison
- *Physical data base design:* data base administration, application designer
- *Process modeling:* data administration, application analyst, user liaison
- *Physical process design:* application designer, data base administration

The assignment of responsibilities shown here points out that applications has given up much of its traditional set of responsibilities. This is quite natural because the applications organization traditionally has a narrow focus (i.e., the views of a single user). Design responsibility has been shifted to the organizational units with a global outlook, such as data administration or data base administration.

RATE OF ORGANIZATIONAL CHANGE

Given that the organization must change to remain efficient, what is the proper rate of change? If an organization attempts to change too rapidly, resistance tends to polarize itself. Furthermore, where change comes quickly there usually is a shortage of design skills.

On the other hand, when change is wrought too slowly, there is a loss of momentum. In a slowly changing environment, one form of resistance may be in the usage of delay tactics. So management must weigh the factors of speed of change —going too fast or too slow—and guide the organization on a consciously chosen course, with a clearly defined long-term goal punctuated by a series of short-term tasks.

EXERCISES

1. Why does modeling usually entail organizational change?
2. Why does modeling require the global perspective? Is there something inherently wrong with the local perspective?
3. How effective can modeling be if it is not committed to *and* understood by management? What if management understands only the buzzwords?
4. (a) What is meant by "long-term commitment" to modeling?
   (b) Where is discipline going to be required?
   (c) Who is going to have to enforce discipline?
5. (a) Why does resistance to change focus on the user and applications? Why is resistance strongest there?
   (b) Is that the only place where there is resistance?
   (c) What can be done to minimize resistance?
6. Why does modeling and integration ultimately boil down to money? How much money is at stake?
7. Why would an organization wish to do things that were not in its best interest? Why would an organization wish to remain immature?
8. Why does modeling address issues of domain and responsibility?

14

Existing Systems and Data Processing Models

In the ideal case integrated systems are built from scratch. But the reality of the matter is that existing systems are, for the most part unintegrated. The irony of integration is that an organization does not recognize the benefits of integration until the organization is forced into the realization by the economic forces that shape the destiny of the company. But the very forces that push a shop into realizing the benefits of integration create a barrier to the full enjoyment of the benefits. The usual progression is that the many unintegrated systems a shop has force the shop into a realization of the value of integration. But those same existing systems hinder the progress a shop can make toward achieving an integrated environment. Existing unintegrated systems need to be changed (usually scrapped), but no manager likes to do away with an existing functional system, whatever its deficiencies.

In the long term, eventual replacement of existing unintegrated systems is the only real solution. In the short term, there is the option to extract data into the modeled form, or if possible, use the existing system in place. Rarely can existing systems be modified into the properly integrated form. One of the fallacies of extracting data from one system to another is that the extracted form of the data has many of the same deficiencies as the original form. In addition, an extract is expensive to write and requires machine resources for execution. The concurrency of the data is an additional limiting factor.

An example of extracting data from existing systems is shown in Fig. 14.1. In the example, four existing systems are extracted into a single unified parts system. However, the integrated parts system exhibits many limitations, such as being only as complete as the existing data, not being current, and so on.

The consideration of performance and availability are less than trivial. There is

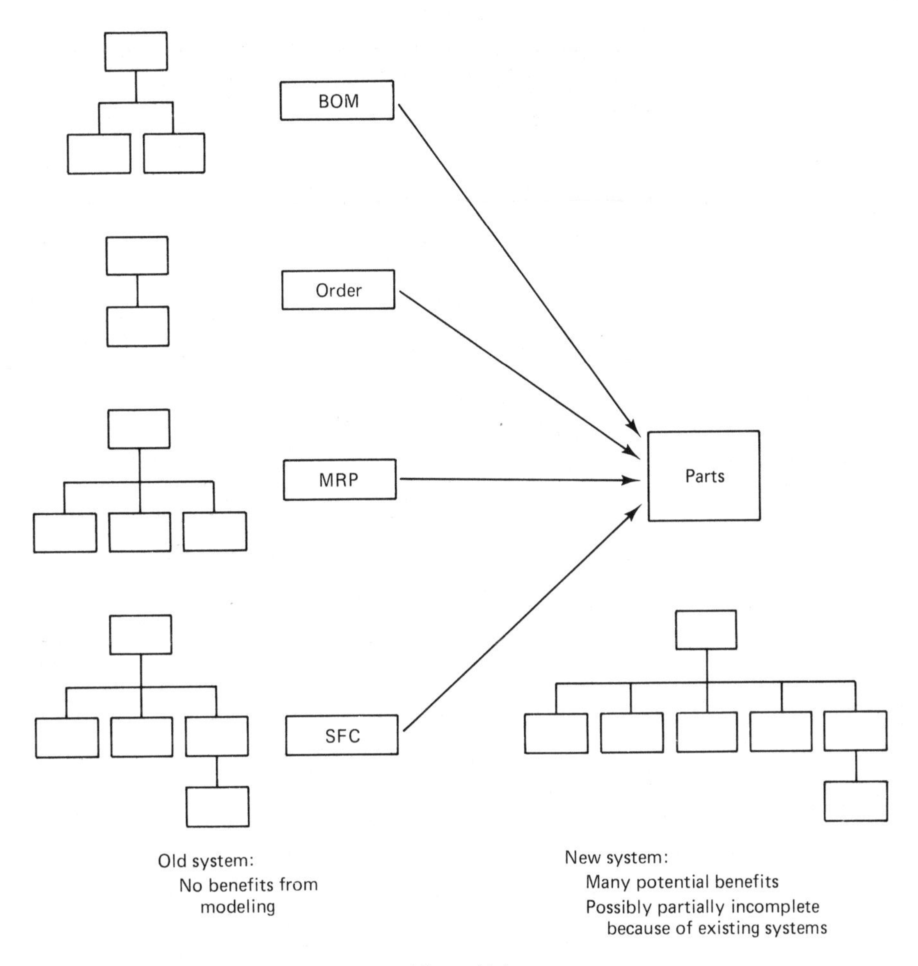

Figure 14.1

much more total work to be done by the system when a strategy of extraction is adopted, and there is more contention for data. From the standpoint of availability, when the integrated data base goes down, many functions of the system come to a halt.

PHASING-IN SYSTEMS

If extraction has many pitfalls and data can be used in place only rarely, what are the other options? The most widely chosen option is to phase the system model in, a piece at a time. This is a fairly easy and natural thing to do. The conceptual model can be realized in many forms.

Complicating the phasing-in of systems is the shift in modes of operation —batch to on-line, operational to decision support, and so on. The process of phasing-in of a system from existing systems is illustrated in Fig. 14.2. This figure shows that four airplane assemblies that currently exist—the F11, P10, B52, and the F85—have been modeled. How they are to be phased in is illustrated in Fig. 14.3. In the first stage, P10 parts are converted to the newly modeled form. Then B52 parts and F85 parts are converted. Last, F11 parts are converted. The phasing could have occurred over a month's time or over several years. During the time between stage I and stage II, part of the data modeled existed in the modeled form and part of the data existed in the original form. Such a migration does not contradict the goals of integration even though ideally the integration model would be created all at once as soon as possible.

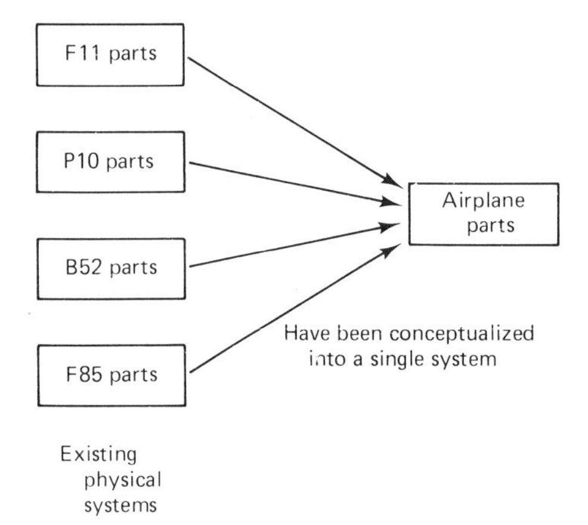

Figure 14.2 Conceptualized system.

Figure 14.3

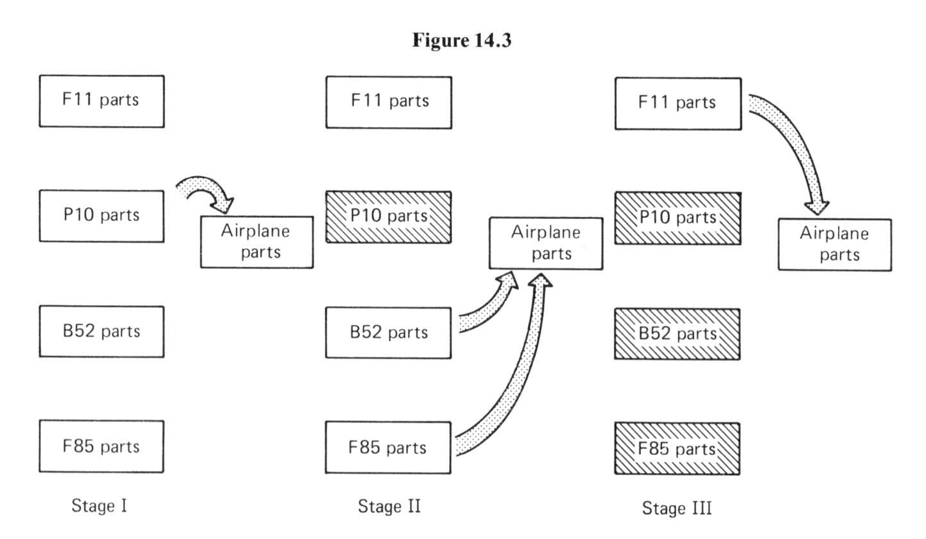

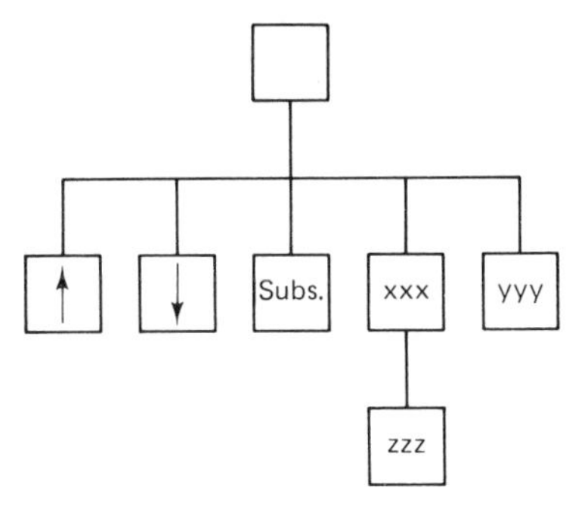

Figure 14.4 Conceptual model.

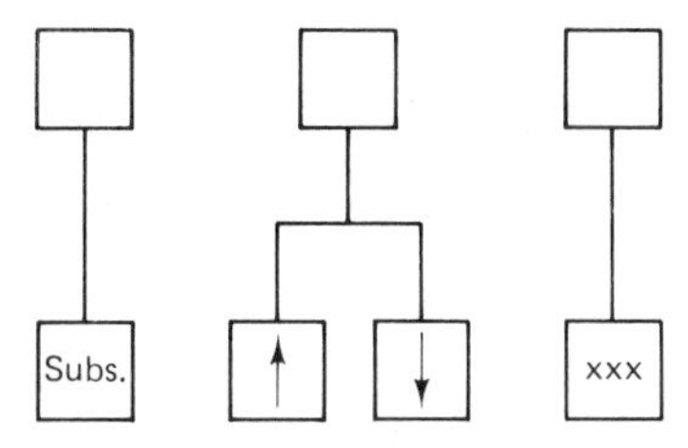

1. Ensure key and existence integrity at the root level.
2. Physical model satisfies most of conceptual model.

Figure 14.5 Existing physical model.

USING DATA IN PLACE

A rare case occurs when data can be used in place, that is, when existing systems can be used to satisfy the data model. Usually, only part of the modeled requirements of the systems can be satisfied and thus this solution represents a temporary solution. Consider the example shown in Fig. 14.4. The model is physically realized by the existence of three physically separate data bases, which collectively make up a significant part of the model's requirements, as shown in Fig 14.5.

IN SUMMARY

Integration poses a dilemma in that by the time the value of integration becomes apparent, a shop has many existing unintegrated systems. Converting existing systems is undesirable because of the cost of conversion. Extracting data into the modeled form is a possibility. However, extraction causes its own development headaches and is undesirable from the standpoint of performance and availability. Phasing-in an existing system is easy to do and often fits with a shop's development schedule. Finally, the cases where data can be used in place are rare.

EXERCISES

1. (a) Is it mandatory that a shop have existing unintegrated systems before it begins to build integrated systems?
   (b) Why does it nearly always happen that a shop has many unintegrated systems?
   (c) How could this paradox be resolved?
2. (a) Why is extraction into the modeled form such an unattractive option?
   (b) Is it *always* unattractive?
   (c) What conditions make it attractive?
3. (a) Why are rewrites universally unpopular?
   (b) Is there such a thing as a "gradual" rewrite?
   (c) Is rewriting a system a form of maintenance or development?
   (d) What is the long-term cost of *not* rewriting a system that needs to be rewritten?
   (e) Does rewriting a system imply that the second version will be any better than the first?
4. (a) Why is phasing-in a system attractive?
   (b) What are the dangers in phasing-in a system?
   (c) Why does phasing-in require stern long-term commitment?
   (d) What costs are there associated with phasing-in a system?
5. Why is using a system in place rarely an option?

15

Documenting Data and Processing Models

Documentation in the integrated environment takes on a new level of importance from its traditional role in data processing. It simply is not adequate to keep some program listings as the primary source of system documentation.

The move to integration requires a long-term commitment to a plan. The plan can be likened to a blueprint and needs to be as well documented as a blueprint. Because the plan must be referred to over a period of time, it must be well documented. Each phase will require a verification of and alignment with the original plan. If for no other reason than adherence to a long-term plan, documentation is vital to the integrated environment.

Another aspect of the integrated environment requiring more than traditional documentation is the reuse of code and data. Without documentation, reuse of code and data is practically impossible in light of even the best design practices. Documentation is one of the cornerstones of the preparation for future needs.

DOCUMENTATION REQUIREMENTS

Documentation of the integrated environment, in general, should be based on the following guidelines:

- *Concurrency:* Documentation must be up to date. This requires more than a casual amount of overhead.
- *Easy access:* Even the best documentation that cannot be retrieved is worthless. Documentation should be publicly available and easy to retrieve.

- *Easy to understand:* Documentation should be written with its future use and readibility in mind, not with the deadline of finishing a project as the driving force. For this reason documentation in the integrated environment should be specifically and separately budgeted.
- *Organized:* A major part of comprehensibility is organization. The reader of documentation should be able to locate relevant information with a minimum of fuss.
- *Concise:* Documentation should say all that is needed to be said, but no more.
- *Clear:* The intent and meaning should be apparent. Simple language with a minimum of mnemonics is desired.
- *Indexed:* An index greatly enhances a reader's chances of locating relevant portions of the document.
- *Relevant:* All aspects of the system needed to describe the system model should be included. A later portion of this chapter suggests what those aspects of the system model might be.

In creating adequate documentation it is most helpful to have a formal, interactive data dictionary. The last choice as a vehicle for documentation is a formal passive data dictionary or an informal data dictionary (although for some parts of the system an informal data dictionary may suffice).

WHAT TO DOCUMENT

The system model has two primary parts: a data model and a process model, as well as a scope. The specific parts of the system model that need documentation are (at least):

- Scope of integration
- Business decomposition
- ERD, including entity definitions (global and local)
- DIS model
- Physical system model (including data characteristics)
- Basic process model
- Functional model
- Detailed process model

For data, entities, attributes, keys, and definitions should be included (at a minimum). For processes, program specifications, subroutine references, a verbal description of the function, inputs/outputs, where used/who used by, and any other idiosyncracies are appropriate to document. In addition, the relationship between processes at each different level is appropriate:

- ERD and BPM
- DIS and function model
- Physical model and DPM

It is clear from the above that program listings by themselves are inadequate.

Other documentation that is useful includes:

- *Abstraction information:* This consists of an outline of the chain of abstraction underlying the ERD.
- *Domain and responsibility definitions:* Origination, design, maintenance, and environmental information is useful. A shop attempting to approach the integrated environment *must* be willing to make the concomitant investment in a documentation. To be effective, documentation *must* occupy a position of importance. If a shop is to succeed, documentation cannot become a passive, after-the-fact, back-room exercise that is sometimes done.

EXERCISES

1. (a) Why is documentation more important to integrated systems than to unintegrated systems?
   (b) Is it possible to achieve long-term goals without very good documentation?
2. What happens when documentation is not kept up to date?
3. (a) What should be documented in an integrated environment?
   (b) Why are program listings and source code definitions woefully inadequate?

16

Decision-Support Systems and Data and Processing Models

Operational systems are those that *run* the company. In operational systems there is much concern for the details, timeliness, and accuracy of data. As an example, consider a bank teller support system. The balance of Ms. Smith needs to be up to the minute, accurate, and available for Ms. Smith to use when she does business with the bank. Ms. Smith does not like to be unable to withdraw her money because of problems with the bank's operational systems.

Decision-support systems are those that are used to *manage* the company. In decision-support systems there is a concern with being able to analyze data in a very flexible fashion. Instead of an emphasis on details, timeliness, and accuracy, the emphasis is on flexibility and the ease of use of the system. As an example, consider a decision-support system for the vice-president of a bank. The data presented to the vice-president is at the summary level. For the purposes of trend analysis, the vice-president could not care less about the accuracy or timeliness of Ms. Smith's account, since her balance will be lumped in with many other balances. The primary motivation for system modeling and building integrated systems is to reduce development and maintenance costs. This motivation applies (although not totally) to the operational environment. However, there are powerful motivations for creating a system model for the decision-support environment as well. Operational systems feed decision-support systems, as shown in Fig. 16.1. Without operational systems that are solidly constructed, there is no basis for decision-support systems. Whereas operational systems can exist without decision-support systems, decision-support systems cannot effectively exist without operational systems.

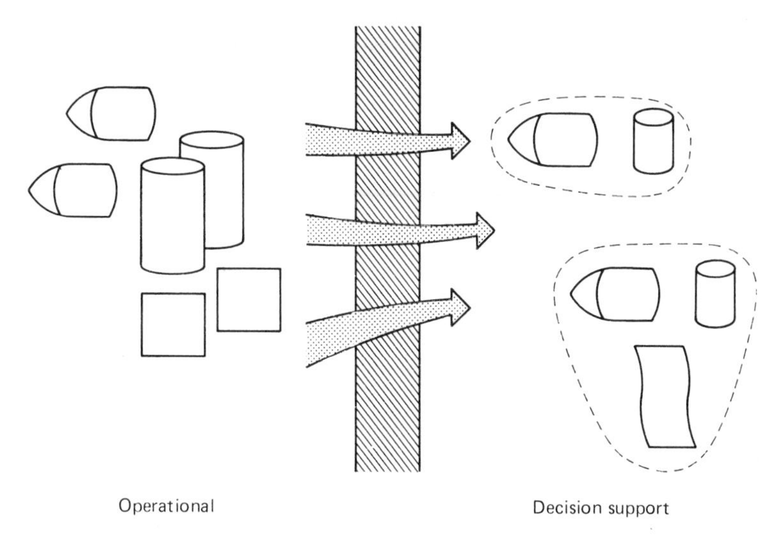

Figure 16.1

SYSTEM MODELING AND DECISION-SUPPORT SYSTEMS

Decision-support systems relate to operational systems primarily by the passing of data from the operational environment. Even though processing (and the manifestation of processing—programs) is important in relating the two environments, processing requirements are an indirect factor, while data is of primary importance. Why, then, is the system model important in the decision-support environment? The need for control of data (performance, availability) is not nearly as high in the decision-support environment as in the operational. But data integrity is equally important in both environments. In the decision-support environment typical processing is summarization of data, factoring of data, selection of data, and so on. For this type of processing to be meaningful, the integrity of the data *must* be paramount. If the base data on which the decision-support system operates is not valid, the resulting decisions must also be invalid.

How, then, do system modeling, integration, and the decision-support environments relate? Consider the environment depicted in Fig. 16.2. In this figure, five unintegrated data bases in the operational environment feed five separate decision-support systems. The data in the operational environment is redundant physically and at the definition level, and in no way integrated. As the data passes from the operational to the decision-support environment, the primary issue is the consistency of the data. The data is inconsistent in the decision-support environment. Different decisions will be made that are supposedly based on the same data. In this case the problems of unintegration are passed on from one environment to the next.

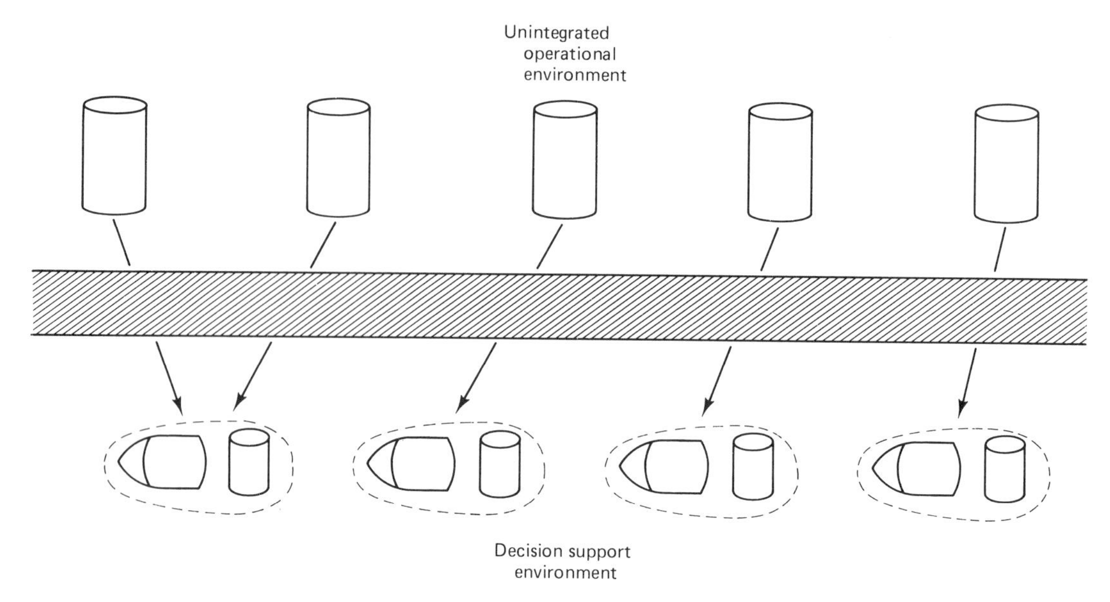

Figure 16.2

The environment depicted in Fig. 16.2 illustrates only the simplest case, however. Consider the much more typical environment depicted in Fig. 16.3. The same problems exist with the unintegrated environment, except that this environment is much more complex. The inconsistency of the data is multiplied, and if properly

Figure 16.3

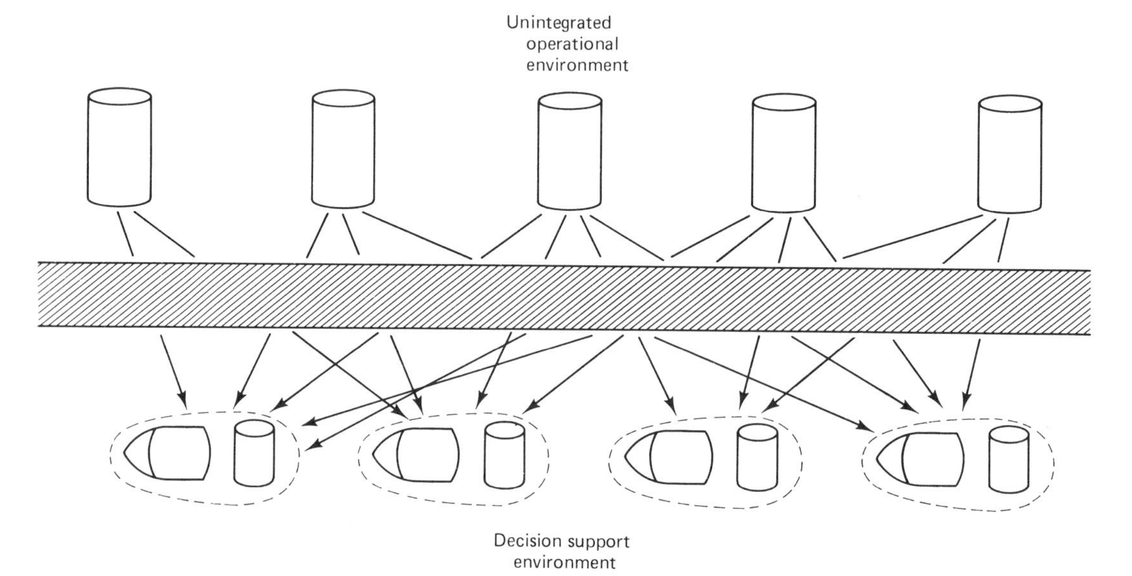

based decisions are difficult in the simple environment depicted, they are practically impossible in the complex environment.

In an attempt to manage the complexities and inconsistencies found in Fig. 16.3, another approach is to attempt to unify the data found in the unintegrated environment, as depicted in Fig. 16.4. In this figure data unified *before* the interface to the decision support environment is created. Although this solves some of the problems, there are *many* problems that remain unsolved and a few new problems that arise, as well. One problem is in the resolution of redundancy. What happens when element A is found in three separate operational data bases? Which data base contributes the data to the centralized interface file? What happens when the form of data elements in one operational data base are not the same as the same elements in another operational data base? What happens when two data bases have a similarly named element that, in fact, is not the same? Or not the same in all cases? And so on.

In addition to requiring more than a few resources to execute, the centralized approach to integrating the two environments is fraught with problems. The centralized interface amounts to an attempt at retrofitting integration onto an unintegrated environment. The results will be fruitless. Another difficulty with the centralized in-

Figure 16.4

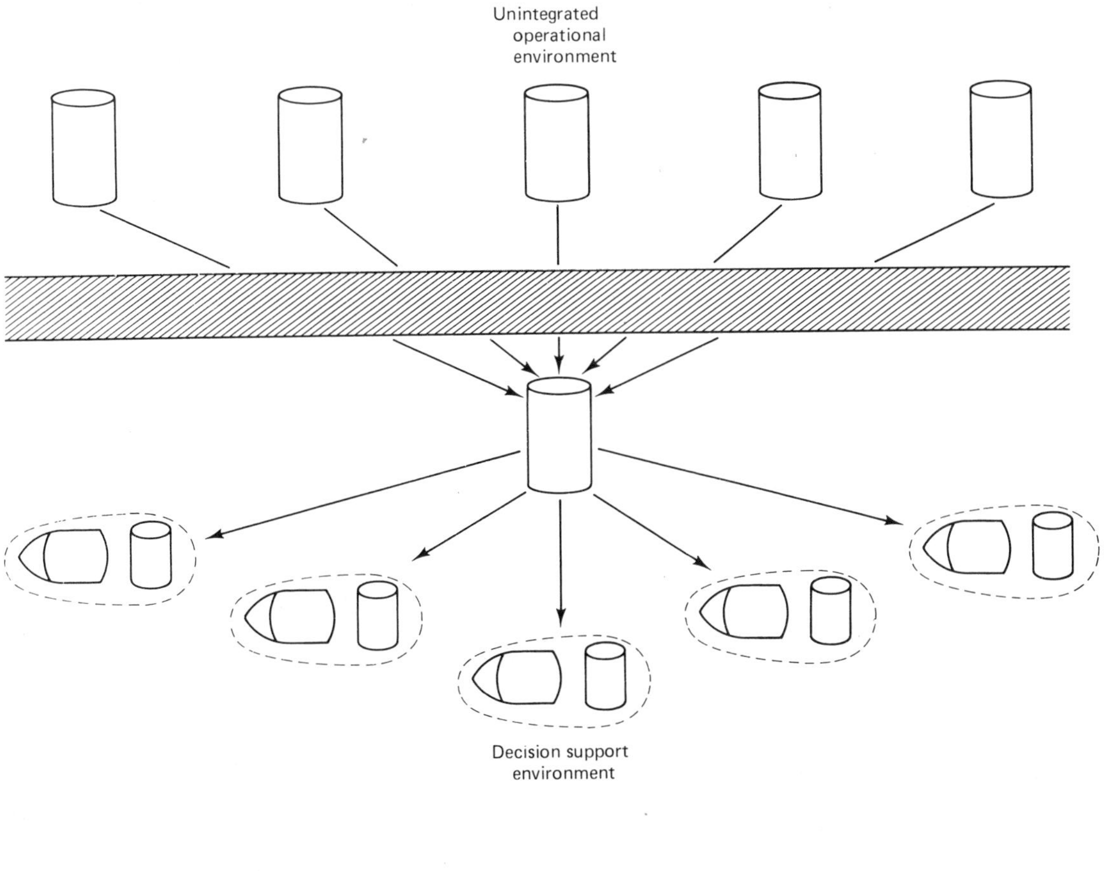

terface is that the interface itself can become a bottleneck in the face of much decision-support activity. There is the additional problem of constructing and maintaining the interface. It then appears that an unintegrated operational environment does not lend itself to an effective decision-support environment.

Consider the decision-support environment in the face of properly modeled, properly integrated environment. Such an environment is shown in Fig. 16.5. In this environment a model for orders and parts has been created. The decision-support environment operates on one or both data bases. There is no question as to the consistency and integrity of the data.

Furthermore, the system model is used to point out where the appropriate data for any given process is located. This environment solves most of the problems discussed previously. However, a new set of problems arises, as shown in Fig. 16.6, where it is seen that many decision support environments are attempting to access parts data. This is a normal phenomenon because integration centralizes data, and as it centralizes, there is a corresponding greater demand to get to the data. In Fig. 16.7 it is seen that an interface between parts and the decision-support environment is created, thus alleviating the performance bottleneck on the operational data base.

Figure 16.5

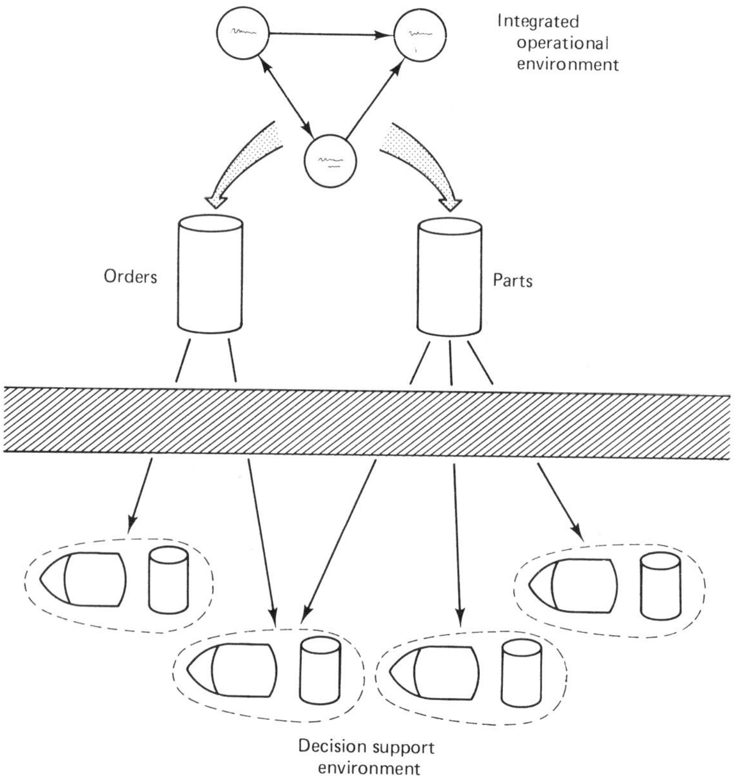

Integrated operational environment

Orders

Parts

Decision support environment

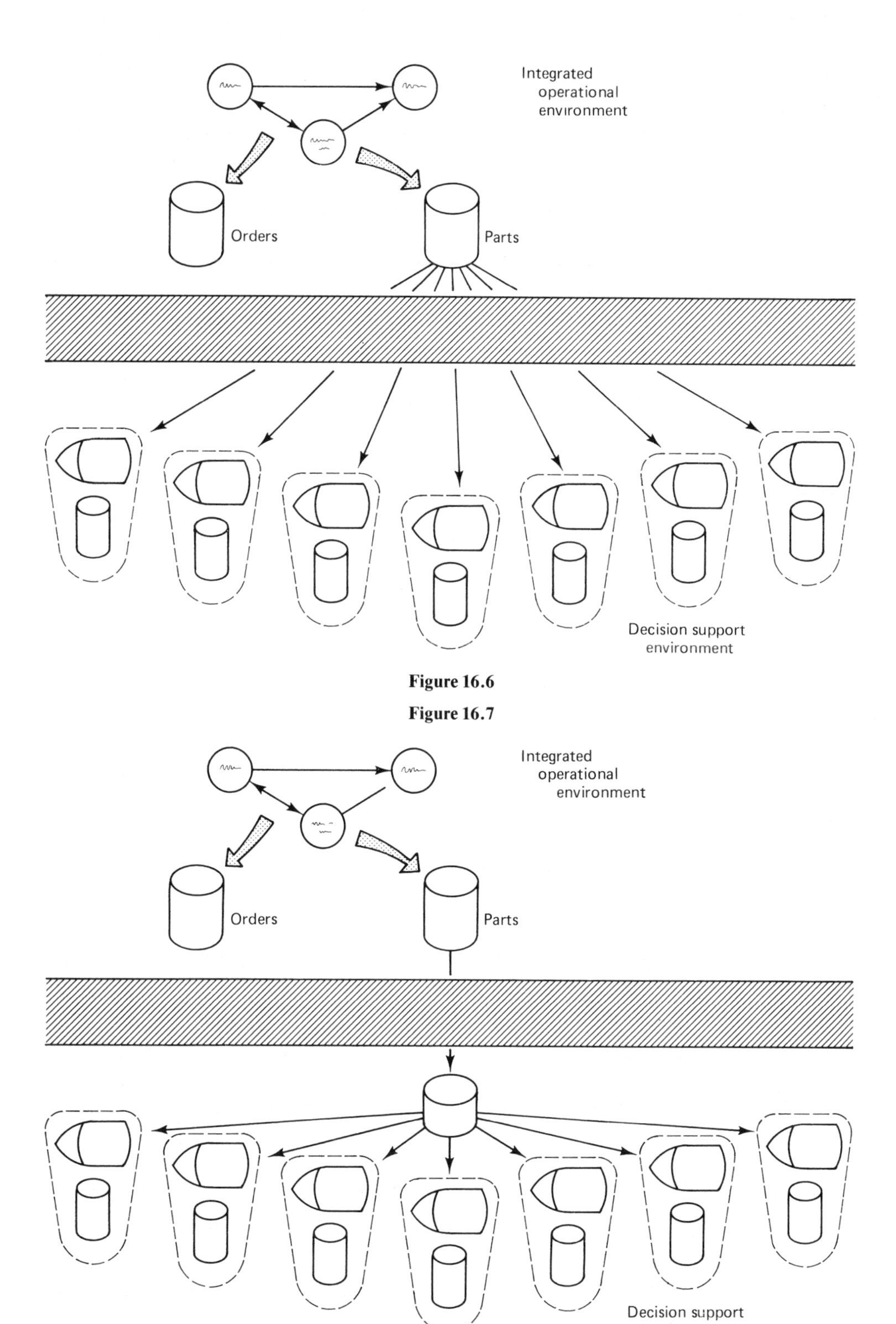

Figure 16.6

Figure 16.7

Note that this interface *does not* have the same problems as those seen in the case of the interface developed for unintegrated systems.

There is a special case of interfacing integrated data bases that are physically split, as shown in Fig. 16.8. In this case an interface is mandatory, as it would be foolish to allow access to one physical data base (e.g., parts 000 to 333) and not to all data bases. One of the difficulties in modelling for the decision support environment and the operational environment is that decision support often looks at data in a summary manner, and operational systems look at data in a detailed manner. Consequently, decision-support models tend to be simplistic and subset oriented, while operation models tend to be very detailed. Yet the two models are intimately tied together. But if the two types of models are included together, the result is often a very complex, very voluminous model which is difficult to use. To resolve this conflict, two (or more) models can be created, one for each level, with notes produced showing the relationship between different entities in each model. The creation of multiple models then produces a simplified, effective system model.

Figure 16.8

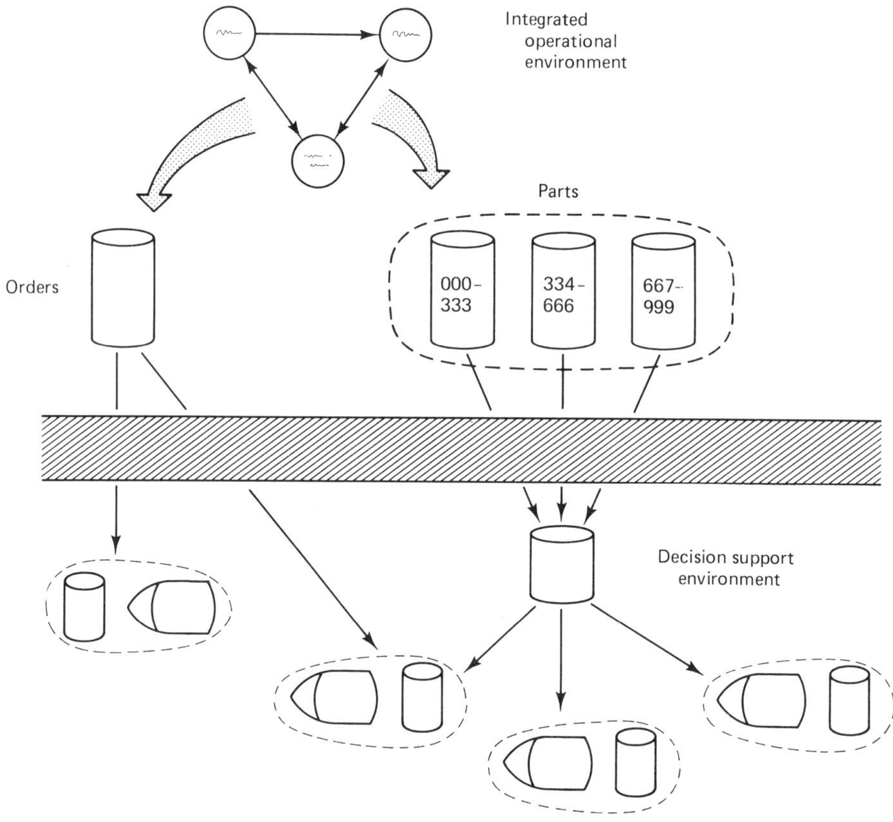

Integrated operational environment

Parts

Orders

000–333

334–666

667–999

Decision support environment

IN SUMMARY

Even though the issues addressed by integration are those of development and maintenance and apply primarily to the operational environment, system modeling applies as well to the decision-support environment. A properly modeled system and the ensuing high degree of integration ensure that data flowing into the decision-support environment are consistent and have a high degree of integrity.

EXERCISES

1. (a) Why is system modeling more important in the operational environment than in the decision-support environment?
   (b) Is modeling unimportant in the decision-support environment?

2. (a) What problems are associated with an unintegrated operational environment and a decision-support environment? Why is the unintegrated operational environment inappropriate to decision-support systems?
   (b) What can be done about unintegrated operational systems? How effectively?

3. (a) What is a centralized interface?
   (b) When can contention become a problem?
   (c) Describe what the centralized interface looks like in terms of:
      (1) An unintegrated operational environment
      (2) An integrated operational environment
      (3) A physically split data base
   What are the problems? How can they be addressed?

4. Data flows from the operational environment to the decision-support environment.
   (a) Doesn't this violate many of the tenets of modeling?
   (b) Doesn't this introduce redundancy at the source level?
   (c) Why aren't programs passed from one environment to the next?

Glossary

Abstraction: The process of categorizing data elements according to common criteria.

Archival mode of operation: The mode of operation in which data stored for historical and future unknown analytical processing after the normal operational life cycle of the data has transpired.

Attribute: An element or data item that is not a unique key.

Bottom-up: The approach to modeling going from the detailed level to a higher, more abstract level; as opposed to top-down.

BPM (basic processing model): The high-level process model; the counterpart in data modeling is the ERD; from the BPM the FM is derived.

Business cycle: A series of business activities, ordered by their occurrence, with a single beginning and ending.

Chandelier perspective: The three-dimensional viewpoint of integrated systems.

Connector: A logical construct used to show a relationship between data elements, data structures, and so on; may be implemented hierarchically, relationally, by a network, and so on.

Consolidated process model: The resulting process model produced by merging common functions found in the various DPMs that represent different dimensions; the consolidated process model is created immediately prior to bottom-up verification of the process model.

DASD: Disk storage; may be hard or floppy.

Data item: The lowest unit of information relating directly to the user's environment. Same as "element."

DBMS (data base management system): System software used to control data bases.

Decomposition: The process of taking an abstraction and breaking it down into its component parts; typical decompositions are done for large functions, organizations, and so on.

Decomposition: The process of taking an abstraction and breaking it down into its component parts; typical decompositions are done for large functions, organizations, and so on.

Dimension: A single view of data as perceived by someone transacting business in the user's environment; a dimension is made up of different user views; there is always a primary business-based dimension; there may be other types of dimensions.

DIS (data item set): The midlevel of the data model—includes elements, keys; is developed from an entity from the ERD; is used as input into the physical model.

DPM (detailed processing model): The lowest level of process modeling; is developed from the FM and consolidated process model; program specifications can be developed from the DPM.

DSS (decision support system): The mode of processing used to manage the company; as opposed to operational systems.

DV (data view): The detailed bottom-up view of a user that serves to verify and complete the detail of the process model.

Elasticity: The ability or inability of a system to be changed with ease—includes data and process considerations.

Element: The lowest unit of information relating directly to the user's environment. Same as "data item."

Entity: Something about which information needs to be recorded; a representation of data at the highest level of abstraction.

Entity dimension map: A diagram (or other representation) of all the dimensions in which a single entity participates.

ERD: A collection of related entities that belong to the user's environment.

Flat file: A collection of data in which data are structured end to end, that is, laid out so that every occurrence of data is unrelated to any other occurrence of data, except by order of occurrence.

FM (functional model): Outline of the function achieved by one of the major processes identified in the basic processing model; is developed from the BPM, serves as input to the consolidated process model.

FOR loop: A logical construct indicating iterative processing.

Functional decomposition: The breakdown of a function into its subfunctions.

Global approach: The perspective looking at the total set of system requirements, each within their own place; as opposed to the local approach.

Global ERD map: A diagram or representation of the various ERDs that belong to a scope of integration.

Global scope of integration: A diagram of the various scopes of integration that belong to a business.

Implementation: The process of translating a conceptual design into a physical system.

Information engineering: The body of thought and practices relating to the development of stable data and process structures.

I/O: Input/output operation; nearly all DBMSs are physically limited by how much I/O a processor can do.

"Is a type of": A logical construct used in abstraction showing how data or processes at one level are included in the category of data or processes at a higher level.

Key: A data element used to identify a collection of data; a key may be unique or nonunique.

Local approach: A perspective in looking at system requirements that considers only the most immediate set of needs; as opposed to the global approach.

Mode of operation: A generic way of executing programs, that is, all programs operating in the same mode share certain characteristics; typical modes of operation are the operational mode, the archival mode, and the decision-support mode.

Object code: Program instruction translated from a high-level language ready to execute in a machine; typically, COBOL is compiled into object code, for example.

Occurrence: A single embodiment of a type of data, of a process, and so on; if a data type were a part number, the data for part number ABC is an occurrence.

Operational systems: Systems used to run the company; typically large, transaction-oriented systems with large data bases where data are stored at a very detailed level; as opposed to decision-support systems.

Performance: In on-line systems, the amount of time from when the enter key is depressed until the first of the response is returned.

Physical characteristics: Information about data other than keys, elements, and the organization of keys and elements; typical physical characteristics include volume of data, volatility of data, and so on.

Prerequisite data: Data that must be present for a process to execute; used in logical process modeling; as opposed to transformation data.

Primary business-based dimension: The view of a system looking at the most basic and most important components of the business of the enterprise.

Processes: Activities; functions; at the programming level, programs.

PV (process view): The view of processes from an end user's perspective; part of a user view; the counterpart is a DV; used in bottom-up verification of the process model.

Recursion: The definition of something in terms of itself; typical recursive structures include bill of materials, family trees, organization charts, and so on.

Redundancy: A multiple occurrence of the same thing.

Relationship: An association between two entities.

Scope of integration: The limitation of a system model.

Semantic redundancy: The redundancy of the definition of data; as opposed to physical redundancy.

Source code: The language actually written by the programmer; source code is compiled or interpreted into a lower form that is executable by a machine.

Top-down: The approach to design that requires first a look at the broad perspective of a system, working successively to a lower and lower level of detail; as opposed to bottom-up.

Transformation data: Data that has the potential to be changed by a process; as opposed to prerequisite data.

User view: The system view from a single user's perspective; made up of data views and process views.

References

BROWN, R. G., *Data Modelling in a Business Environment,* Macmillan, New York, 1983.

BROWN, R. G., "Logical Data Base Design Techniques," The Database Design Group, Mountainview, Calif., 1981.

BROWN, R. G., "Structured Database Design—An Overview," Guide 54, Anaheim, Calif., May 12, 1982.

BROWN, R. G., and PARKER, D. STOTT, "LAURA: A Formal Data Model and Her Logical Design Methodology," Very Large Data Base Conference, Italy, 1983.

DATE, C. J., *An Introduction to Database Systems,* Addison-Wesley, Reading, Mass., 1975.

INMON, W. H., *Effective Data Base Design,* Prentice-Hall, Inc., Englewood Cliffs, N.J., 1981.

INMON, W. H., *Integrating Data Processing Systems: In Theory and in Practice,* Prentice-Hall, Englewood Cliffs, N.J., 1984.

KROENKE, D., *Database Processing,* Science Research Associates, Palo Alto, Calif.

MARTIN, J., *Computer Data-Base Organization,* 2nd ed., Prentice-Hall, Englewood Cliffs, N.J., 1977.

SMITH, J. M., and SMITH, D. C. P., "Conceptual Database Design," INFOTECH, State of the Art Report on Database Design, 1980.

SMITH, J. M., and SMITH, D. C. P., "Database Abstractions: Aggregation," *Communications of the ACM, 20*:6, June 1977.

SMITH, J. M., and SMITH, D. C. P., "Database Abstractions: Aggregation and Generalization," *ACM Transactions on Database Systems, 2*:2, June 1977, 105–133.

TSCHRITZSIS, D. C., and LOCHOVSKY, F. H., *Data Models,* Prentice-Hall, Englewood Cliffs, N.J., 1981.

ULLMAN, J. D., *Principles of Database Systems,* Computer Science Press, Potomac, Md., 1980.

Index

H

I

J

K

L

M

N

O

P

Q

R

S

T

U

V

W